PE EXAM PREP

CIVIL ENGINEERING
PE SAMPLE EXAM

James H. Banks, PhD
San Diego State University

Braja M. Das, PhD, PE
California State University–Sacramento (Retired)

Bruce E. Larock, PhD, PE
University of California–Davis (Emeritus)

Javier Rodríguez Mejías, PE
University of Phoenix–Guaynabo

Gregory Mills, PE
Western Kentucky University

Thomas B. Nelson, PhD, PE
University of Wisconsin–Platteville

Philip J. Parker, PhD, PE
University of Wisconsin–Platteville

Ben J. Stuart, PhD, PE
Ohio University

Alan Williams, PhD, SE, Chartered Eng.
California Department of Transportation

Kenneth J. Williamson, PhD, PE
Oregon State University

This publication is designed to provide accurate and authoritative information in regard to the subject matter covered. It is sold with the understanding that the publisher is not engaged in rendering legal, accounting, or other professional service. If legal advice or other expert assistance is required, the services of a competent professional person should be sought.

President: Mehul Patel
Vice President & General Manager: David Dufresne
Vice President of Product Development and Publishing: Evan M. Butterfield
Editorial Project Manager: Laurie McGuire
Director of Production: Daniel Frey
Production Editor: Caitlin Ostrow
Production Artist: Cepheus Edmondson
Creative Director: Lucy Jenkins

Copyright 2008 by Dearborn Financial Publishing, Inc.®

Published by Kaplan AEC Education
30 South Wacker Drive
Chicago, IL 60606-7481
(312) 836-4400
www.kaplanaecengineering.com

All rights reserved. The text of this publication, or any part thereof, may not be reproduced in any manner whatsoever without written permission from the publisher.

Printed in the United States of America.

08 09 10   10 9 8 7 6 5 4 3 2 1

CONTENTS

Introduction v

CHAPTER 1 Morning Exam 1
PROBLEMS 1

CHAPTER 2 Afternoon Exam—Environmental Engineering and Water Resources 15
PROBLEMS 15

CHAPTER 3 Afternoon Exam—Geotechnical Engineering 25
PROBLEMS 25

CHAPTER 4 Afternoon Exam—Structural Engineering 39
PROBLEMS 39

CHAPTER 5 Afternoon Exam—Transportation Engineering 55
PROBLEMS 55

CHAPTER 6 Afternoon Exam—Construction Engineering 67
PROBLEMS 67

CHAPTER 7 Solutions 83
SOLUTIONS TO CHAPTER 1 PROBLEMS 83
SOLUTIONS TO CHAPTER 2 PROBLEMS 97
SOLUTIONS TO CHAPTER 3 PROBLEMS 112
SOLUTIONS TO CHAPTER 4 PROBLEMS 126
SOLUTIONS TO CHAPTER 5 PROBLEMS 141
SOLUTIONS TO CHAPTER 6 PROBLEMS 149
SOLUTION SUMMARIES 165

Introduction

OUTLINE

HOW TO USE THIS BOOK V

BECOMING A PROFESSIONAL ENGINEER VI
Education ■ Fundamentals of Engineering/Engineer-in-Training (FE/EIT) Exam ■ Experience ■ Professional Engineer Exam

CIVIL ENGINEERING PROFESSIONAL ENGINEER EXAM VII
Examination Development ■ Examination Structure ■ Design Standards ■ Exam Dates ■ Exam Procedure ■ Exam-Taking Suggestions ■ Exam Day Preparations ■ What to Take to the Exam ■ Examination Scoring and Results ■ Acknowledgments ■ Errata

HOW TO USE THIS BOOK

Civil Engineering PE Sample Exam gives you an opportunity to simulate the experience of taking the Professional Engineer (PE) exam in civil engineering. It is hoped that this experience will make you feel more relaxed and prepared on the day of the actual exam. The problems and solutions covered in this book will also give you a good review of concepts and analytical techniques that are likely to appear on the exam. A good approach to using this book for optimal exam preparation follows:

1. Set aside a 4-hour block of time to answer the 40 questions in Chapter 1. Don't look at the solutions until you have answered all the questions. Use the reference texts you plan to bring to the actual exam as needed to help you choose the correct answer to each question.

2. Choose from Chapters 2–6 the one afternoon depth exam that you plan to take at the actual exam and set aside another 4-hour block of time to answer those 40 questions. Follow the same procedures as for Chapter 1. To closely approximate the exam experience, you may want to tackle the morning exam and your chosen afternoon exam consecutively on the same day, with just an hour-long break between them. This is the format of the actual PE exam.

3. When you have completed a chapter, turn to Chapter 7 to review the correct answers and detailed solutions. Make note of which topic areas gave you the most difficulty—these are areas you may want to focus on further as you review for the exam.

Some of the problems in the sample exams may be more complex than any you are likely to encounter on the actual exam. Where this occurs, the goal is to give you a good review of a concept or technique that may be important on the exam. The problem and its detailed solution function as a focused review of a relevant topic. These rigorous problems should help you become more than adequately prepared for the exam day.

For a more thorough review, you may want to use this book in combination with its companion texts: *Civil Engineering PE License Review* and *Civil Engineering PE Problems & Solutions*. *Civil Engineering PE License Review* provides a conceptual review of key terms and equations, analytical methods, and design considerations, including solved examples. *Civil Engineering PE Problems & Solutions* provides extensive opportunity to practice solving problems; it is organized in the same topic order as the *License Review* book so that you can refer back and forth between problems and concepts as needed. Together, these three books provide a comprehensive review for the exam.

BECOMING A PROFESSIONAL ENGINEER

There are four distinct steps to achieve registration as a Professional Engineer: (1) education; (2) passing the Fundamentals of Engineering/Engineer-in-Training (FE/EIT) exam; (3) professional experience; and (4) the Professional Engineer (PE) exam, more formally known as the Principles and Practice of Engineering Exam. These steps are described in the following sections.

Education

The obvious appropriate education is a BS degree in civil engineering from an accredited college or university. This is not an absolute requirement. Alternative, but less acceptable, education is a BS degree in a field other than civil engineering, a degree from a nonaccredited institution, or four years of education but no degree.

Fundamentals of Engineering/Engineer-in-Training (FE/EIT) Exam

Most candidates for PE registration are required to take and pass this eight-hour multiple-choice examination. Although different states call it by different names (e.g., Fundamentals of Engineering, Engineer-in-Training, or Intern Engineer), the exam is the same in all states. It is prepared and graded by the National Council of Examiners for Engineering and Surveying (NCEES). Review materials for this exam are found in other books published by Kaplan AEC, such as *Fundamentals of Engineering FE/EIT Exam Preparation*.

Experience

Several years of acceptable experience are typically required before one is permitted to take the Professional Engineer exam. Both the length and character of the experience will be examined.

Professional Engineer Exam

The second national exam is called Principles and Practice of Engineering by the NCEES, but almost everyone else calls it the Professional Engineer, or PE, exam. All U.S. states, the District of Columbia, Guam, and Puerto Rico use the same NCEES exam.

CIVIL ENGINEERING PROFESSIONAL ENGINEER EXAM

Laws regulating the practice of engineering are adopted to protect the public from incompetent practitioners. Most states require engineers who work on projects involving public safety to be registered or to work under the supervision of a registered professional engineer. In addition, many private companies encourage or require engineers in their employ to pursue registration as a matter of professional development. Engineers in private practice and those who wish to consult or serve as expert witnesses typically also must be registered. There is no national registration law; registration is based on individual state laws and is administered by boards of registration in each of the states. You can find a list of contact information for and links to the various state boards of registration at the Kaplan AEC Web site: *www.kaplanaecengineering.com*. This list also shows the exam registration deadline for each state.

Examination Development

Initially, the states wrote their own examinations, but beginning in 1966, the NCEES took over the task for some states. Now the NCEES exams are used by all states. This makes it much easier for an engineer to move from one state to another and achieve registration in the new state.

The development of the engineering exams is the responsibility of the NCEES Committee on Examinations for Professional Engineers. The committee is composed of representatives from industry, consulting, and education, plus consultants and subject matter experts. The starting point for the exam is an engineering task analysis survey that the NCEES does at roughly five- to ten-year intervals. People in industry, consulting, and education are surveyed to determine what civil engineers do and what knowledge is needed to do it. From this data, the NCEES develops what it calls a "matrix of knowledge," which forms the basis for the exam structure described in the next section.

The actual exam questions are prepared by the NCEES committee members, subject matter experts, and other volunteers. All participants must hold professional registration. Using workshop meetings and correspondence by mail, the committee members write the questions and circulate them for review. Although based on an understanding of engineering fundamentals, the problems require the application of practical professional judgment and insight.

Examination Structure

The exam is organized into breadth and depth sections.

The morning breadth exam consists of 40 multiple-choice questions covering the following five areas of civil engineering: environmental engineering and water resources, geotechnical, structural, transportation, and construction. Each topic area is weighted equally, at 20 percent. You will have four hours to complete the breadth exam.

The afternoon depth portion is actually five exams, one on each of the morning breadth topics. You choose the one depth exam you wish to take; the obvious choice is whichever one best matches your training and professional practice. You will have 4 hours to answer the 40 multiple-choice questions that make up the depth exam.

For more information on the topics and subtopics and their relative weights on the breadth and depth portions, visit the NCEES Web site at *www.ncees.org*.

The PE exam is given over two 4-hour sessions, with 40 questions in each session. All questions are multiple-choice with four answer choices.

Design Standards

The PE exam for civil engineering incorporates design standards and codes from several widely used industry references. Candidates can expect to see structural, construction, and transportation engineering questions that require them to apply equations, safety factors, and other criteria from the design standards.

NCEES posts the relevant standards for each exam on its Web site. You should download these lists and become familiar with the organization and application of each reference. You probably will want to bring a copy of relevant sections of each reference to the exam itself.

Exam Dates

The NCEES prepares Professional Engineer exams for administration on one Friday in April and one Friday in October of each year. Some state boards administer the exam twice a year, while others offer the exam once a year. The scheduled exam dates for the next ten years can be found on the NCEES Web site *(www.ncees.org/exams/schedules/)*.

People seeking to take a particular exam must apply to their state board of registration several months in advance.

Exam Procedure

Before the morning four hour session begins, the proctors pass out an exam booklet and a solutions pamphlet to each examinee. The solutions pamphlet contains grid sheets on right-hand pages. Only work on these grid sheets will be graded. The left-hand pages are blank and are for use as scratch paper. The scratch work will *not* be considered in the scoring.

The proctors also will provide each examinee with a mechanical pencil for use in recording answers; this is the only writing instrument allowed. Do not bring your own pencil lead or eraser. If you need an additional pencil during the exam, a proctor will supply one.

If you finish more than 15 minutes early, you may turn in the booklets and leave. If you finish in the last 15 minutes, however, you must remain through the end of the hour to ensure a quiet environment for those still working and an orderly collection of materials.

The afternoon session will begin following a one-hour lunch break. The afternoon exam booklet will be distributed along with an answer sheet.

Exam-Taking Suggestions

Give yourself time to prepare for the exam in a calm and unhurried way. Many candidates like to begin several months before the actual exam. Target a number of hours per day or week that you will study and reserve blocks of time for doing so. Creating a review schedule on a topic-by-topic basis is a good idea. Remember to allow time for both reviewing concepts and solving practice problems.

In addition to review work that you do on your own, you may want to join a study group or take a review course. A group study environment may help you stay committed to a study plan and schedule. Group members can create additional practice problems for one another and share tips and tricks.

You may want to prioritize the time you spend reviewing specific topics according to their relative weight on the exam (as identified by the NCEES) or according to your areas of relative strength and weakness. There may be an exam topic that you have little or no exposure to. This would be a good area to focus on, time permitting, provided you feel strong in other areas.

People familiar with the psychology of exam taking have several suggestions for exam candidates:

- Passing a competency exam involves two skills. One is the skill of illustrating your knowledge. The other is the skill of exam taking. The first may be enhanced by a systematic review of the technical material. Exam-taking skills, on the other hand, may be improved by practicing with problems presented in a format similar to the exam format.

- Because there is no penalty for guessing on the multiple-choice problems, you should answer all of the problems. Even when you are going to guess, however, use a logical approach. Attempt to first eliminate one or two of the four alternatives. If you can do this, the chance of selecting a correct answer obviously improves, from 1 in 4 to 1 in 3 or 1 in 2.

- Plan ahead with a strategy. Which is your strongest area? Can you expect to see several problems in this area? What about your second-strongest area? What is your weakest area?

- Plan ahead with a time allocation. Compute how much time you will allow for each of the major subject areas on the morning exam. You might allocate a little less time per problem for those areas in which you are most proficient, leaving a little more time in subjects that are more difficult for you. Your plan should include reserving a block of time for especially difficult problems, for checking your scoring sheet, and finally for making last-minute guesses on any problems you did not work. Your strategy may also include time allotments for two passes through the exam: the first to work all problems for which answers are obvious to you, the second to return to the more complex, time-consuming problems and the ones at which you may need to guess. A time plan gives you the confidence of being in control and keeps you from making the serious mistake of misallocating time.

- Read all four multiple-choice answers before making a selection. An answer in a multiple-choice question is sometimes a plausible decoy but not really the best answer.

- Do not change an answer unless you are absolutely certain you have made a mistake. Your first reaction is likely to be correct.

- Do not sit next to a friend, a window, or other potential distractions.

Exam Day Preparations

The exam day will be a stressful and tiring one. This will be no day to have unpleasant surprises. For this reason, we suggest making an advance visit to the examination site. Try to determine items such as the following:

- How much time should I allow for travel to the exam on that day? Plan to arrive about 15 minutes early. This way you will have ample time but not too much time. Arriving too early and mingling with others who are anxious will increase your anxiety and nervousness.

- Where will I park?

- How does the exam site look? Will I have ample workspace? Where will I stack my reference materials? Will it be overly bright (bring sunglasses) or cold (bring a sweater), or noisy (bring earplugs)? Would a cushion make the chair more comfortable?

- Where are the drinking fountain and lavatory facilities?

- What about food? Should I bring a snack for energy during the exam? A bag lunch for during the break probably makes sense.

What to Take to the Exam

The NCEES guidelines allow you to bring the following reference materials and aids into the examination room for your personal use only:

1. Handbooks and textbooks, including the applicable design standards

2. Bound reference materials, provided the materials remain bound during the entire examination. The NCEES defines *bound* as books or materials fastened securely in their covers by fasteners that penetrate all papers. Examples are ring binders, spiral binders and notebooks, plastic snap binders, binders with brads or screw posts, and so on.

3. A battery-operated, silent, nonprinting, noncommunicating calculator from the NCEES list of approved calculators. For the most current list, see the NCEES Web site (*www.ncees.org*). You must also determine whether your state permits preprogrammed calculators. Bring extra batteries for your calculator just in case; many people feel that bringing a second calculator is also a very good idea.

At one time, the NCEES prohibited bringing "review publications directed principally toward sample questions and their solutions" into the exam room. This led to restrictions against bringing some kinds of publications to the exam. *State boards may adopt the NCEES guidelines or may adopt either more or less restrictive rules.* Thus, an important step in preparing for the exam is to know what will—and will not—be permitted at your exam location. We recommend that you obtain a written copy of your state's policy for the specific exam you will be taking. Occasionally, there has been confusion at individual examination sites, so a copy of the exact applicable policy will not only enable you to prepare your materials carefully and correctly but will also ensure that the exam proctors allow all proper materials that you bring to the exam.

As a general rule, we recommend that you plan well in advance what books and other materials to take to the exam. Obtain them promptly so you use the same materials in your review that you will use in the exam.

License Review Books

The review books you use to prepare for the exam are good choices to bring to the exam itself. After weeks or months of studying, you will be very familiar with their organization and content, so you'll be able to locate quickly the material you want to reference during the exam. Keep in mind the caveat just discussed—some state boards will not permit you to bring review books that consist largely of sample questions and answers into the exam room.

Textbooks

If you still have your college or university textbooks, they are the ones you should use in the exam, unless they are too out-of-date. To a great extent, these books will be like old friends with familiar notation.

Bound Reference Materials

The NCEES guidelines suggest that you can take to the exam any reference materials you wish, so long as you prepare them properly. You could, for example, prepare several volumes of bound reference materials with each volume covering a particular category of problem. Use tabs so specific material can be located quickly. If you do a careful and systematic review and prepare a lot of well-organized materials, you just may find that you are so well prepared that you will not have left anything of value at home.

Other Items

In addition to the reference materials just mentioned, you should consider bringing the following to the exam:

- Clock—You must have a time plan and a clock or wristwatch.

- Exam assignment documents—Take along the letter assigning you to the exam at the specified location. To prove you are the correct person, also bring at least one photo ID (such as a driver's license).

- Items suggested by advance visit—If you visit the exam site, you will probably think of an item or two to add to your list.

- Clothing—Plan to wear comfortable clothes. You probably will do better if you are slightly cool.

- Box for everything—You need to be able to carry all your materials to the exam and have them conveniently organized at your side. A cardboard box is probably the answer.

Examination Scoring and Results

The questions are machine scored by scanning. The answers sheets are checked for errors by computer. Marking two answers to a question, for example, will be detected, and no credit will be given.

Your state board will notify you whether you have passed or failed roughly three months after the exam. Candidates who do not pass the exam the first time may take it again. If you do not pass, you will receive a report listing the percentages of questions you answered correctly for each topic area. This information can help focus the review efforts of candidates who need to retake the exam.

The PE exam is challenging, but analysis of previous pass rates shows that the majority of candidates do pass it the first time. By reviewing appropriate concepts and practicing with exam-style problems, you can be in that majority. Good luck!

Acknowledgments

Several contributors provided review and updating of material in this book. The publisher is grateful to:

- David Fanella, PhD, SE
- Denise D. Gravitt, PhD, CIT, CDS, Western Kentucky University
- Ganesh Thiagarajan, PhD, PE, University of Missouri at Kansas City

Errata

The author and publisher of this book have been careful to avoid errors, employing technical reviewers, copy editors, and proofreaders to ensure the material is as flawless as possible. Any known errata and corrections are posted on the product page at our Web site, *www.kaplanaecengineering.com*. If you believe you have discovered an inaccuracy, please notify the engineering editor at Kaplan AEC Education:

Engineeringpress@kaplan.com
Fax: 312-836-9958
Kaplan AEC Education
30 S. Wacker Drive, Suite 2500
Chicago, IL 60606

CHAPTER 1

Morning Exam

1.1 A reinforced concrete beam is simply supported over a span of 10 feet.

Design data:
Normal weight concrete, $f'_c = 3000$ psi
Reinforcing steel = Grade 60, ASTM A-615
Beam width = 12 in.
Effective depth = 18 in.

The strength design provisions of the ACI *Building Code Requirements for Standard Concrete*, ACI 318-05, apply.

The beam supports a live load of 7 kips/ft and a dead load (including its own weight) of 3 kips/ft. At midspan, the number of No. 8 bars required to resist the applied moment is most nearly:
a. 6
b. 5
c. 4
d. 3

1.2 Exhibit 1.2 shows the force acting on column AB of a frame with sidesway uninhibited. This force is caused by gravity loads and includes the self-weight of the members.

Design data:
Steel members to ASTM A992, $F_y = 50,000$ psi
Columns = W14 × 53

$$A = 15.6 \text{ in.}^2$$
$$I_x = 541 \text{ in.}^4$$
$$r_x = 5.89 \text{ in.}$$
$$r_y = 1.92 \text{ in.}$$

Beams = W18 × 40

$$I_x = 612 \text{ in.}^4$$

AISC *Steel Construction Manual*, 13th ed. (2006), applies.

You may select **either** the ASD **or** the LRFD option.

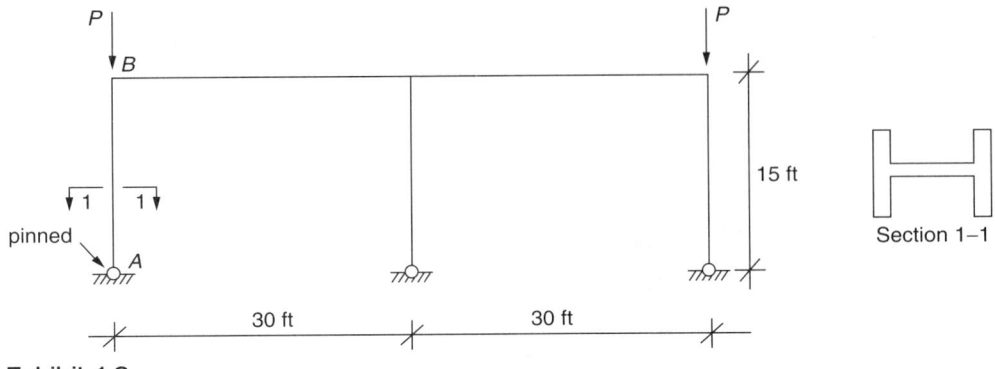

Exhibit 1.2

ASD Option

The allowable load *P* (kips) that, in the absence of bending moment, may be applied to the column is most nearly:
a. 244
b. 246
c. 248
d. 250

LRFD Option

The design axial strength of the column *P* (kips) in the absence of bending moment is most nearly:
a. 367
b. 369
c. 370
d. 373

1.3 A wood-framed workshop is shown in Exhibit 1.3.
The ICC *2006 International Building Code* and the following data apply:

Roof framing members = 2 × Douglas Fir–Larch select structural
Moisture content < 19%
Plywood roof diaphragm = 15/32-inch Structural I

The diaphragm is blocked at plywood panel edges, and seismic effects govern. The seismic design load in the north-south direction at the roof diaphragm level is 280 plf.

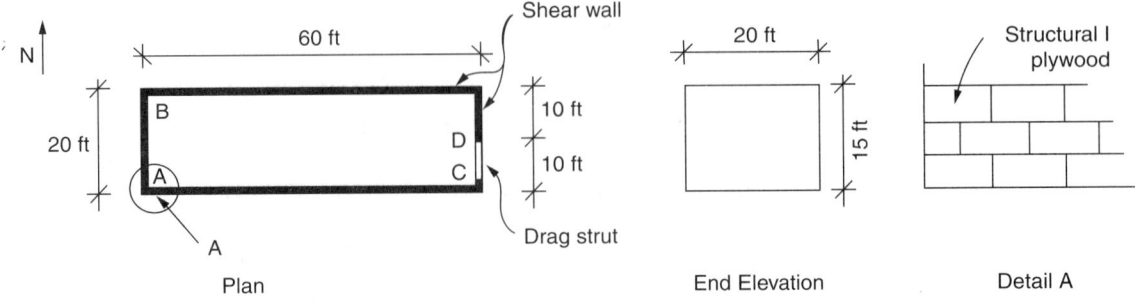

Exhibit 1.3

Using 10d common nails with $1\tfrac{5}{8}$-inch penetration, the nail spacing, in inches, required along the diaphragm boundary AB is most nearly:
a. 2
b. $2\tfrac{1}{2}$
c. 4
d. 6

1.4 The reinforced concrete bridge shown in Exhibit 1.4 is simply supported over an effective span of 50 feet. AASHTO *Standard Specifications for Highway Bridges,* 17th ed. (2002), applies.

The design loading is HS20-44 truck loading.

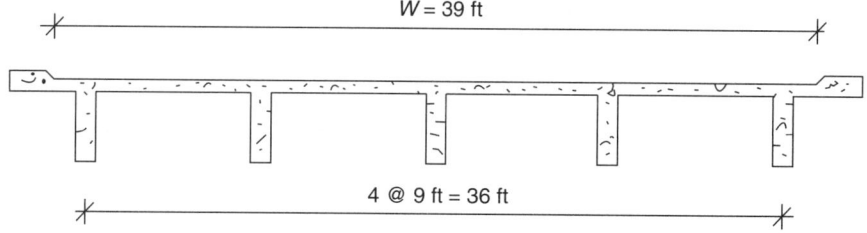

Exhibit 1.4

The maximum live-load moment (kip-ft) in an interior beam, including impact, is most nearly:
a. 600
b. 700
c. 800
d. 900

1.5 The concrete gravity dam shown in Exhibit 1.5 retains water at the level shown. Due to seepage under the dam, hydrostatic uplift pressure develops under the base and varies from a maximum of the full water head at point X to zero at the toe. The coefficient of friction between the concrete dam and the foundation material is 0.5.

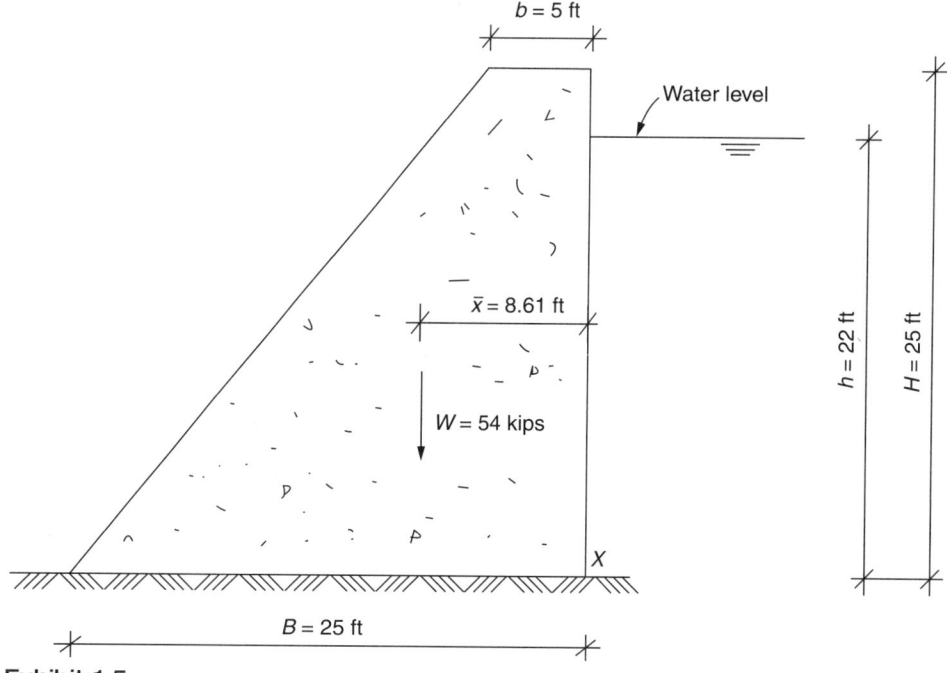

Exhibit 1.5

The factor of safety against overturning due to all loads and the uplift pressure is most nearly:
a. 2.0
b. 2.1
c. 2.2
d. 2.3

1.6 An office building located on a site with an undetermined soil profile is a bearing wall structure constructed with special reinforced masonry shear walls. The maximum considered earthquake response acceleration parameters are $S_s = 0.75g$ and $S_1 = 0.50g$. The natural period of vibration of the building is $T_a = 0.15$ second.

The ICC *2006 International Building Code* and the 2005 ASCE/SEI 7 *Minimum Design Loads for Buildings and Other Structures* apply.

The value of the seismic response coefficient C_s is most nearly:
a. 0.16
b. 0.14
c. 0.12
d. 0.10

1.7 A nominal 24-inch-square, solid-grouted, concrete block masonry column has a height of 20 feet and may be considered pinned at each end.

Codes:
ACI 530-05, *Building Code Requirements for Masonry Structures*

Materials:
Masonry strength, $f'_m = 1500$ psi
Modulus of elasticity = 1000 ksi
Reinforcing steel = Grade 60, ASTM A-615
Reinforcement = Eight No. 8 bars

The axial load (kips) that the column can support, including its own weight, using Allowable Stress Design, is most nearly:
a. 260
b. 270
c. 280
d. 290

1.8 A reinforced concrete beam is simply supported over a span of 10 feet.

Design data:
Normal weight concrete, $f'_c = 3000$ psi
Reinforcing steel = Grade 60, ASTM A-615
Beam width = 12 in.
Effective depth = 18 in.

The strength design provisions of the ACI *Building Code Requirements for Structural Concrete*, ACI 318-05, apply.

The beam supports a live load of 7 kips/ft and a dead load (including its own weight) of 3 kips/ft. The maximum shear force V_u, in kips, at the critical section is most nearly:
a. 50
b. 55
c. 60
d. 65

1.9 The concrete-lined open channel ($n = 0.013$) shown in Exhibit 1.9 has streamwise slope of 0.25 percent.

The discharge in this channel is most nearly:
a. 100 m³/s
b. 10 m³/s
c. 8 m³/s
d. 5 m³/s

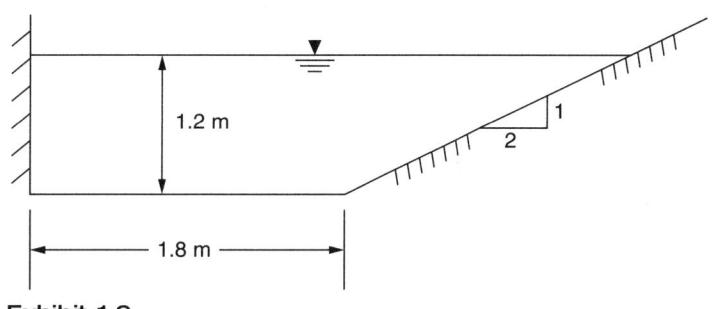

Exhibit 1.9

1.10 The pipes ABCDEF in Exhibit 1.10 are the mains for a small new subdivision. To test this system for adequacy for fire flow at point F, with the supply coming from pipe AB, it is desired to replace lines BCDF and BEF with one 8-in. line that will be hydraulically equivalent to the two proposed lines. Assume that the Hazen-Williams formula applies.

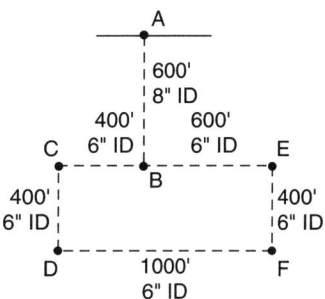

Exhibit 1.10

The length of the equivalent 8-in. line is most nearly:
a. 375 ft
b. 625 ft
c. 1000 ft
d. 1475 ft

1.11 A rectangular open channel has a rough concrete lining ($n = 0.017$) and conveys a discharge of 22.0 m³/s. If the channel bottom slope is 0.01, then the hydraulically most efficient cross-sectional dimensions, including 0.5 m freeboard, are most nearly:
a. 1.5 m high and 3.0 m wide
b. 2.0 m high and 3.5 m wide
c. 2.0 m high and 3.0 m wide
d. 3.0 m high and 2.0 m wide

1.12 The rectangular swimming pool at a high school is 75 ft long and 60 ft wide and varies from 4 ft to 6 ft in depth. It is suspected of having a leak in the bottom, so data in the table below were collected for one week.

Day	Precipitation P, in.	Evaporation E, in.	WSE, ft
0			250.00
1		0.4	
2		0.4	
3	0.85		
4		0.5	
5	0.30		
6		0.5	
7		0.6	249.25

The water surface elevation, WSE, is recorded at the end of each day, and the other values are daily data. Based on these data, the size of the leak, in gallons/day, is most nearly:
a. 1000
b. 2000
c. 3000
d. 3500

1.13 A 60-acre parcel of single-family residences has a time of concentration of 45 minutes. If this area is subjected to 1.2 inches of rainfall in an hour, a reasonable estimate of the maximum expected peak runoff, in ft³/s, is most nearly:
a. 18
b. 24
c. 30
d. 36

1.14 A confined aquifer with a transmissibility $T = 40$ m²/day lies under some farmland. If the aquifer were developed, the slope of the piezometric surface would be 0.25 m/km. The amount of water per day, in m³, that would flow through a 1-km-wide section of this aquifer is most nearly:
a. 0.25
b. 1.0
c. 10.0
d. 100.0

1.15 A water has the following composition:

Ca^{2+}, 71 mg/L; Mg^{2+}, 19 mg/L; Na^+, 11 mg/L; HCO_3^-, 120 mg/L; SO_4^{2-}, 101 mg/L; Cl^-, 55 mg/L.

The pH is 7.5.

Based on the given information, this water is:
a. negatively charged
b. approximately neutrally charged
c. positively charged
d. Cannot be determined from the information given

1.16 A river has a flow rate of 1 m³/s, a dissolved oxygen concentration of 8 mg/L, no significant BOD, and a temperature of 20°C. A wastewater treatment plant discharges to the river at a flow of 0.1 m³/s with a BOD_5 of 20 mg/L and a DO of 1.5 mg/L.

The initial deficit after the waste is mixed in the river is about:
a. 7 mg/L
b. 2 mg/L
c. Depends on the temperature of the river and waste
d. b and c

1.17 A primary sedimentation basin for a municipal wastewater treatment plant treats a flow of 3 MGD and removes 70 percent of the total suspended solids. The basin is 60 ft in diameter with a depth of 15 ft.

The solids flux in this basin is about:
a. 1 lb/ft²-d
b. 2 lb/ft²-d
c. Independent of detention time
d. Independent of depth

1.18 The TSS removed in the basin in Problem 1.17 would be about:
a. 100 kg
b. 1000 kg
c. 2000 kg
d. 4000 kg

1.19 An activated sludge plant is operated as shown in Exhibit 1.19:

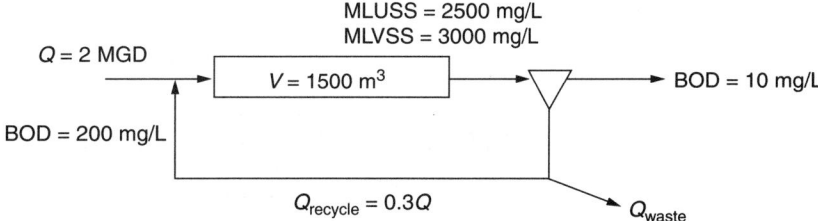

Exhibit 1.19

If the solids retention time is 5 days, the sludge waste per day is about:
a. 10 kg/d
b. 100 kg/d
c. 1000 kg/d
d. 10,000 kg/d

1.20 The observed yield value for the scenario in Problem 1.19 is:
a. about 0.1 mg TUSS/mg BOD
b. about 0.5 mg TUSS/mg BOD
c. about 1.0 mg TUSS/mg BOD
d. Cannot be determined from the data given

1.21 Disinfection with chlorine can be modeled as a first-order reaction. A reactor has a flow rate of 12,000 liters per hour, and the desired removal of organisms is from 10×10^6/100 mL to less than 1/100 mL. The decay coefficient is 0.35/hr.

Short-circuiting in this complete mixed reactor will:
a. reduce treatment efficiency
b. increase the required volume
c. make the reactor operate more like a plug flow reactor
d. a and b

1.22 The volume of the reactor described in Problem 1.21 required for a plug flow condition is:
a. less than 10^5/L
b. less than 10^6/L
c. independent of k_1
d. independent of Q

1.23 A sludge from a municipal wastewater treatment plant is 67 percent volatile and 5 percent solids.

The specific gravity of the sludge is:
a. 0.99
b. 1.00
c. 1.01
d. 1.02

1.24 If digestion destroys 50 percent of the volatile solids, the density of the sludge would be about:
a. 1.2
b. 1.3
c. 1.4
d. 1.5

1.25 A compacted soil sample has the following properties:

Volume, $V = 0.15$ m^3
Moisture content, $w = 11\%$
Moist weight, $W = 2.63$ kN
Specific gravity of soil solids, $G_s = 2.67$

The void ratio, e, of the soil is nearly equal to:
a. 0.42
b. 0.55
c. 0.66
d. 0.82

1.26 The minimum and maximum void ratios of a sand are 0.47 and 0.8, respectively. The specific gravity of the soil solids, G_s, is 2.65. A direct shear test was conducted on this soil (dry). For the test,

Void ratio, $e = 0.62$
Specimen size = $2 \times 2 \times 1.5$ in. (height)
Normal load, $N = 40$ lb
Shearing load of failure, $S = 28$ lb

The soil friction angle for the shear test is nearly:
a. 30°
b. 35°
c. 40°
d. 45°

1.27 The minimum and maximum void ratios of a sand are 0.45 and 0.82, respectively. The relative density of this sand in the field is 45.9 percent. The specific gravity of the soil solids, G_s, is 2.66.

The dry unit weight of the sand in the field is:
a. 85.2 lb/ft^3
b. 100.6 lb/ft^3
c. 110.1 lb/ft^3
d. 115.6 lb/ft^3

1.28 Following are the laboratory results of a Proctor compaction test. The specific gravity of the soil solids, G_s, is 2.75. Volume of compaction mold = 1/30 ft^3.

Moisture Content, %	Moist Unit Weight of Compaction, lb/ft^3
12	113.7
14	123.1
16	128.1
18	131.6
20	131.4
22	130.6

The maximum dry unit weight of compaction is nearly equal to:
a. 105 lb/ft^3
b. 112 lb/ft^3
c. 117 lb/ft^3
d. 125 lb/ft^3

1.29 A soil profile is shown in Exhibit 1.29.

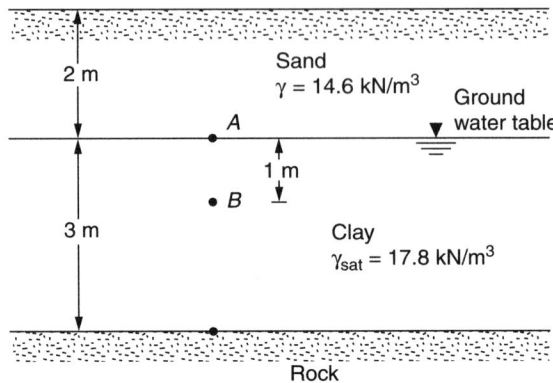

Exhibit 1.29

The effective vertical stress, σ', at B will be about:
a. 28 kN/m²
b. 37 kN/m²
c. 74 kN/m²
d. 85 kN/m²

1.30 A flow net for a sheet pile cofferdam is shown in Exhibit 1.30. For the soil, the hydraulic conductivity is 2.5×10^{-5} ft/min.

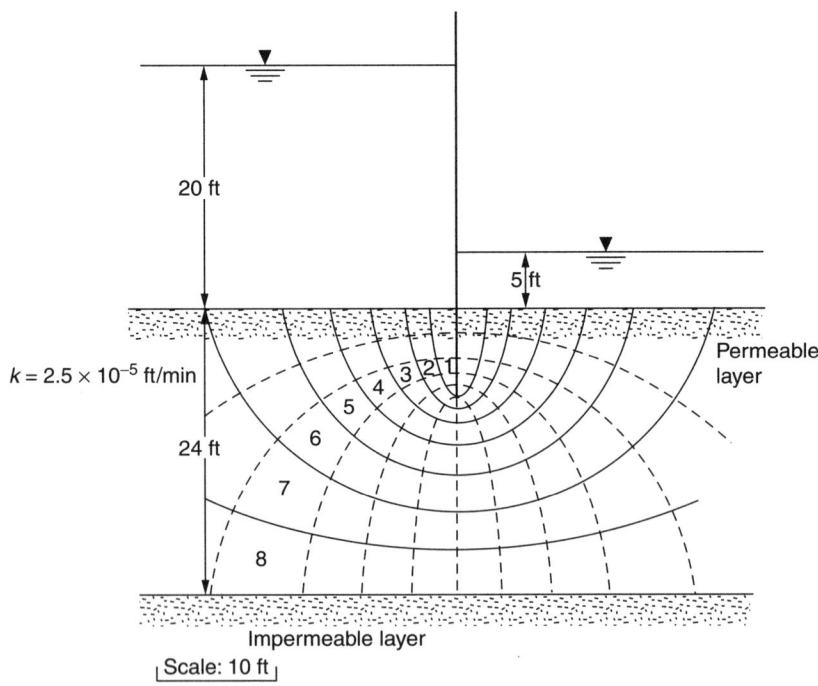

Exhibit 1.30

The seepage through the permeable layer is nearly:
a. 5×10^{-5} ft³/min/ft
b. 10×10^{-5} ft³/min/ft
c. 15×10^{-5} ft³/min/ft
d. 25×10^{-5} ft³/min/ft

1.31 A shallow square foundation is shown in Exhibit 1.31. The size of the foundation is $B \times B$ in plan.

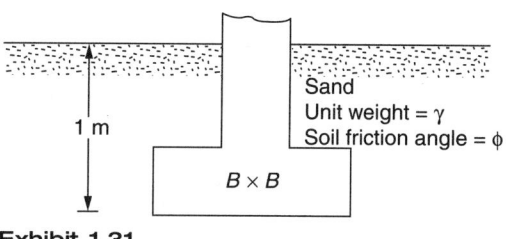

Exhibit 1.31

Given: $B = 1.5$ m, $\gamma = 16$ kN/m³, $\phi = 35°$. The gross ultimate vertical load-bearing capacity of the foundation, q_u, will be approximately equal to:
a. 550 kN/m²
b. 1100 kN/m²
c. 1450 kN/m²
d. 2000 kN/m²

1.32 Drained triaxial test results of undisturbed core samples from the backfill of a retaining wall (Exhibit 1.32) are as follows:

Test No. 1: $\sigma_3 = 19$ lb/in.²; $\sigma_1 = 69$ lb/in.²
Test No. 2: $\sigma_3 = 25$ lb/in.²; $\sigma_1 = 91$ lb/in.²

In situ soil properties: Void ratio, $e = 0.50$
Specific gravity, $G_s = 2.65$
Moisture content, $w = 10.0\%$

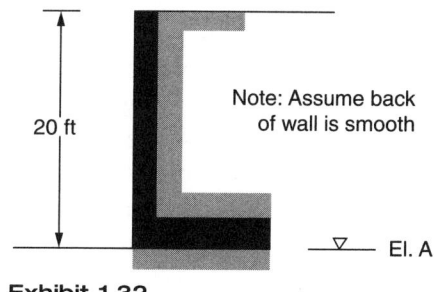

Exhibit 1.32

The Rankine active force behind the wall with the groundwater level at Elevation A is nearly equal to:
a. 5000 lb/ft
b. 6600 lb/ft
c. 7700 lb/ft
d. 3000 lb/ft

1.33 A rural freeway has two lanes in the direction of travel being analyzed. It has a measured free-flow speed of 70 mph. The heavy vehicle factor is estimated to be 0.87, the peak hour factor to be 0.90, and the driver population factor to be 0.92. The maximum hourly volume is expected to be 2700 vehicles per hour. The level of service is:
a. B
b. C
c. D
d. E

1.34 The following data apply to the westbound approach of an isolated intersection with a pretimed signal:

Cycle = 60 s
Saturation flow rate (for two lanes) = 3300 veh/h
Saturation (v/c) ratio = 0.85
Effective green = 22 s

Control delay is to be measured over a 15-minute interval. There is no queue present at the beginning of the analysis interval. The level of service is:
a. A
b. B
c. C
d. D

1.35 The following table gives earthwork cross-sectional areas for three successive locations on a roadway:

	Earthwork Cross Section, ft²	
Station	Cut	Fill
10 + 00	25.0	0.0
10 + 40	15.0	12.0
10 + 95	0.0	20.0

If material shrinks by 20 percent (that is, volume of a unit mass is 20 percent less in a compacted fill than in the material's existing state), the net waste or borrow for this job is:
a. waste 3.4 yd³ (volume material occupies in existing state)
b. borrow 6.7 yd³ (volume material occupies in compacted fill)
c. waste 9.8 yd³ (volume material occupies in existing state)
d. borrow 5.6 yd³ (volume material occupies in compacted fill)

1.36 The stopping sight distance for a vehicle traveling at 70 km/h on a +3 percent grade is most nearly:
a. 60 m
b. 70 m
c. 80 m
d. 100 m

1.37 The PI of the 1200 ft horizontal curve shown in Exhibit 1.37 is located 1322 ft from the PT of the preceding curve, which is located at station 5 + 43.

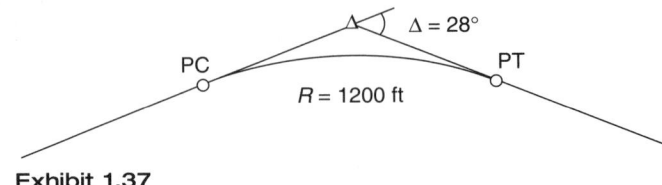

Exhibit 1.37

The station of the PT of the curve shown is most nearly
a. 21 + 65
b. 20 + 50
c. 22 + 20
d. 21 + 50

1.38 A two-lane highway (with one 3.6 m lane in each direction) has a horizontal curve with a radius of 400 m. An obstruction to vision is located 8.2 m from the edge of traveled way on the inside of the curve. The stopping sight distance is most nearly:
a. 150 m
b. 160 m
c. 180 m
d. 220 m

1.39 A vertical curve on a rural highway joins a +2.5 percent grade to a −3.5 percent grade. The design speed is 55 mph. The minimum length of vertical curve needed to provide stopping sight distance is most nearly:
a. 600 ft
b. 700 ft
c. 800 ft
d. 900 ft

1.40 A 300-meter-long vertical curve joins a −2.3 percent grade to a +3.6 percent grade. The station of the PVC is 7 + 43, and the elevation of the PVC is 87.621. The station and elevation of the low point of the curve are most nearly:
a. station = 8 + 93, elevation = 86.4 m
b. station = 8 + 60, elevation = 86.3 m
c. station = 8 + 93, elevation = 84.2 m
d. station = 8 + 60, elevation = 84.9 m

CHAPTER 2

Afternoon Exam—Environmental Engineering and Water Resources

2.1 A municipal waste is treated with an activated sludge system with recycle after primary clarification. The solids retention time for the reactor depends on:
a. effluent quality
b. recycle TSS concentration
c. hydraulic retention time
d. efficiency of the primary clarifier

2.2 A settling test was conducted on some waste-activated sludge treated in a secondary clarifier. The recycle ratio (Q_r/Q) is 0.3, and the overflow rate is 600 gpd/ft^2. The results of the test are:

MLSS (mg/L)	Setting Velocity (ft/h)
4000	7.8
6000	3.8
8000	1.8
10,000	1.0
12,000	0.2

The underflow rate is:
a. 100 gpd/ft^2
b. 200 gpd/ft^2
c. 300 gpd/ft^2
d. 400 gpd/ft^2

2.3 Given the scenario in Problem 2.2 of a sludge concentration of 6000 mg/L, the downward velocity of a particle is:
a. 1.7 ft/hr
b. 3.8 ft/hr
c. 5.5 ft/hr
d. 7.8 ft/hr

2.4 A trickling filter containing plastic media is to be designed at an organic loading of 2 kg BOD/m^3-d, hydraulic loading of 50 m^3/m^2-d, and a recirculation ratio of 2:1. Assume a municipal waste flow of 1 MGD after clarification.

The area required based on the hydraulic loading would be about:
a. 5 m^2
b. 10 m^2
c. 15 m^2
d. 25 m^2

2.5 The area required based on the organic loading in Problem 2.4 would be about:
a. 5 m^2
b. 10 m^2
c. 15 m^2
d. 25 m^2

2.6 A rotating biological contactor is being used to remove BOD from a municipal wastewater after clarification. To achieve 80 percent BOD removal, the hydraulic loading rate should be less than:
a. 0.1 m^3/m^2-d
b. 0.2 m^3/m^2-d
c. 0.3 m^3/m^2-d
d. 0.4 m^3/m^2-d

2.7 An activated sludge reactor with a solids retention time of nine days is to achieve nitrification. pH control may be necessary because:
a. nitrification produces acid
b. the activated sludge process consumes alkalinity
c. nitrification releases ammonia
d. a and b

2.8 Alum is being dosed to remove 10 mg/L and achieve 90 percent removal of phosphorus. The flow is 5000 m^3/d. If the chemical formula for the alum is Al$_2$(SO$_4$)$_3$ • 18H$_2$O, the amount of alum required is:
a. 150 mg/L
b. 250 mg/L
c. 350 mg/L
d. 500 mg/L

2.9 Primary municipal sludge with a concentration of 4 percent TSS and a flow rate of 50 m^3/d is to be treated by anaerobic digestion. If the solids retention time required is 15 d, the volume is:
a. 500 m^3
b. 750 m^3
c. 1000 m^3
d. 1500 m^3

2.10 A wastewater effluent is treated with chlorine gas (Cl_2) to give a free residual of 0.5 mg/L. The amount of chlorine required is:
a. much greater than 0.5 mg/L
b. dependent on the COD of the effluent
c. dependent on the NH_3–N concentration of the effluent
d. All of the above

2.11 A sludge digester is loaded at a rate of 200 kg COD/d with a solids retention time of 20 days. The waste utilization is about 80 percent. The cell growth in the reactor is about:
a. 5 kg/d
b. 10 kg/d
c. 15 kg/d
d. 20 kg/d

2.12 The quantity of methane produced in Problem 2.11 is about:
a. 1000 ft^3/d
b. 2000 ft^3/d
c. 3000 ft^3/d
d. 4000 ft^3/d

2.13 A conventional (plug flow) activated sludge system is converted to a complete-mix system with the same θ_c and θ_n values. The diffuse aerators in the conventional system, as compared to those in the complete-mix system, will:
a. have greater capacity
b. require great control
c. be spaced differently
d. b and c

Problems 2.14–2.15

A small Midwestern city produces a wastewater flow of 5.5 MGD and has an influent BOD of 170 mg/L. The WWTP consists of an aerated grit chamber (AGC), primary clarifier, activated sludge (AS) tank, final clarifier, and disinfection through UV radiation. The primary removes 60 percent of the solids and 35 percent of the BOD. Laboratory data for the AS process determined a biological yield of 0.6 and a microbial decay constant of 0.06 day^{-1} in the 155,000-ft^3 tank when there was an MLSS concentration of 2100 mg/L. Final plant effluent has a BOD of 10 mg/L. You may assume a flow peaking factor of 3.

2.14 Determine the dimensions (L × W × D) of the AGC if the design detention time is 1 minute at peak flow and the length:width:depth ratios are 1.5:1.5:1.
a. 10.5 × 10.5 × 7 ft
b. 12 × 12 × 8 ft
c. 13.5 × 13.5 × 9 ft
d. 15 × 15 × 10 ft

2.15 Determine the depth of the circular primary clarifier if the overflow rate is 1000 gpd/ft² and the detention time is two hours at Q_{ave}.
a. 10 ft
b. 11 ft
c. 12 ft
d. 13 ft

Problems 2.16 and 2.17
A small river receives a WWTP discharge of 5 MGD with a BOD of 20 mg/L and a DO concentration of 1.0 mg/L. The river is 40 feet wide and flows at 1 ft/s, but depth and temperature vary by season. You may assume that the BOD in the river is 0.9 mg/L and the DO concentration is at 90 percent of the saturation value, which also varies with temperature. The biological activity coefficient $(k_d)_{20}$ is 0.3 day^{-1}, and the stream reaeration coefficient $(k_a)_{20}$ is 0.25 day^{-1}. You may assume temperature correction coefficients for k_d of 1.056 and for k_a of 1.024.

2.16 Determine the minimum DO concentration in the stream in the summer when the river temperature is 22°C and flows at a depth of 3 feet. You may assume a WWTP discharge temperature of 16°C.
a. 7.3 mg/L
b. 7.5 mg/L
c. 7.7 mg/L
d. 7.9 mg/L

2.17 Determine the minimum DO concentration in the stream in the winter when the river temperature is 4°C and flows at a depth of 4 feet. You may assume a WWTP discharge temperature of 13°C.
a. 11.0 mg/L
b. 11.3 mg/L
c. 11.5 mg/L
d. 11.8 mg/L

2.18 Which of the following statements regarding the similarities between types of photosynthetic bacteria is true?
a. Cyanobacteria and purple bacteria are closely related.
b. Cyanobacteria and green bacteria are closely related.
c. Purple bacteria and green bacteria are closely related.
d. There are no similarities among cyanobacteria, purple bacteria, and green bacteria.

2.19 The release of a toxic organic compound from an underground storage tank has just been identified. Initial tests have determined the following:

groundwater table = 12 feet bgs (below ground surface)
confining layer = 40 feet bgs
width of contaminated area = 100 feet
length of contaminated area = 600 feet
average contaminant concentration in the aqueous phase (C_{aq}) = 2.3 mg/L
soil water partition coefficient (K_{SW}) = 0.4
soil porosity = 32%
contaminant specific gravity = 0.87

Observation wells at radii of 30 feet and 40 feet have drawdowns of 4.6 feet and 0.3 foot, respectively, when water is pumped from the extraction well at a rate of 5 gpm. Estimate the soil hydraulic conductivity.
a. 146.4 ft/yr
b. 133.6 ft/yr
c. 119.3 ft/yr
d. 101.8 ft/yr

2.20 Particles exiting a flocculation basin settle out according to Type I settling. The particles have a size range of 10 μm to 85 μm and a specific gravity of 1.8. If the tank is sized to remove 100 percent of the largest flocs, what percentage of the smallest flocs will be removed?
a. 1.4%
b. 12%
c. 14%
d. 21%

2.21 A concrete conduit having a cross-sectional area of 0.90 m² and a wetted perimeter of 3.75 m conveys water at a mean velocity of 2.50 m/s. The smallest head loss that can be expected in 1000 m of this conduit is most nearly:
a. 10.0 m
b. 7.0 m
c. 5.0 m
d. 4.0 m

2.22 A 1000 ft long horizontal pipe with a 12 in. diameter leaves a reservoir with water surface elevation 200 ft at elevation 180 ft. Through an abrupt contraction, this line is connected to a 1000 ft long pipe with a 6 in. diameter running to elevation 100 ft, where it enters a reservoir with water surface elevation 130 ft. Assume a friction factor of 0.02.

The discharge between the reservoirs is most nearly:
a. 2.0 ft³/s
b. 2.2 ft³/s
c. 4.5 ft³/s
d. 5.0 ft³/s

2.23 A rectangular open channel is 10 ft wide and conveys 500 ft³/s. The depth of flow is 3.00 ft immediately upstream of a short, smooth, well-designed channel hump that raises the channel floor 0.5 ft. The depth of flow atop this hump, in feet, is most nearly:
a. 3.00
b. 3.35
c. 3.85
d. 4.25

2.24 A rectangular open channel is 5.00 m wide and is lined with concrete. It conveys 25.0 m³/s of water and has two longitudinal sections separated by a break in slope. The normal depths of flow, far upstream and downstream of the break in slope, are 1.00 m and 1.50 m, respectively. The flow sequence from the initial depth to the final depth is most nearly:
a. a hydraulic jump from 1.00 m to 1.50 m at the break in slope
b. gradually varied flow downstream of the break in slope, followed by a hydraulic jump directly to the 1.50 m depth
c. a hydraulic jump from the 1.00 m depth at the break in slope, followed by gradually varied flow to the 1.50 m depth
d. gradually varied flow from the 1.00 m depth to the 1.50 m depth without a hydraulic jump

2.25 A 300-acre watershed experiences two precipitation events: (i) in 40 minutes, a total of 40 acre-feet of rain is deposited; (ii) in 25 minutes, a rainfall depth of 1.2 in. falls on the watershed.

Of these two events, the larger rainfall intensity, in in./h, is most nearly:
a. 1.2
b. 1.6
c. 2.4
d. 3.0

2.26 Ten hectares of land contain a few neighborhood businesses in one half, and the other half is a park. The time of concentration for runoff from this area is about 20 minutes. The land receives rain for an hour at a rate of 1 cm per 15 minutes. The peak outflow discharge for this area is most nearly:
a. 0.3 m³/s
b. 0.5 m³/s
c. 0.7 m³/s
d. 0.9 m³/s

2.27 The first table on the next page describes the relation between elevation and storage for a reservoir with a spillway outlet crest at elevation 753 ft. Above this elevation, the outflow can be approximated by the expression $Q = 5.0(H - 753)^{3/2}$ ft³/s. Also, given in a second table are inflow data for one week at this reservoir; the reservoir water surface elevation is 750 when inflow begins.

Elev. H, ft	Storage S_{AF}, acre-feet
750	300
751	320
752	340
753	360
754	385
755	410
756	440
757	480

Time, days	1	2	3	4	5	6	7
Average inflow, ft³/s	10	25	35	30	25	20	15

The peak outflow discharge from the reservoir is most nearly:
a. 30.0 ft³/s
b. 27.5 ft³/s
c. 25.0 ft³/s
d. 22.5 ft³/s

2.28 The system served by a pump emptying a lift station can be approximated by this system curve: $h_p = A + B \times Q$, where A and B are constants (10 ft and 2 cfs^{-1}, respectively) and Q is the flow rate. According to the pump curve shown in Exhibit 2.28a, the operating discharge is most nearly:
a. 1 cfs
b. 2 cfs
c. 3 cfs
d. 4 cfs

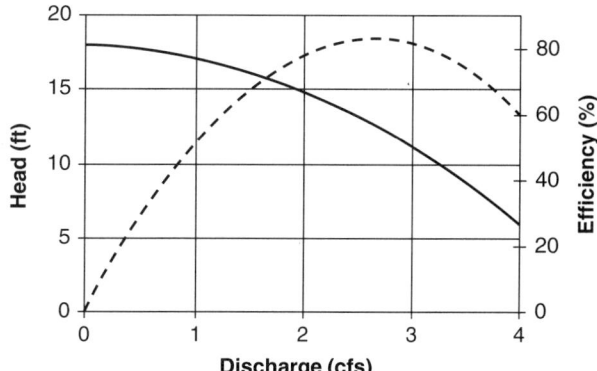

Exhibit 2.28a

2.29 The flow in a 12-inch concrete pipe when it flows full is 1 cfs. The flow and velocity in that same pipe when it flows half full is most nearly:
a. 1 cfs and 1.3 fps
b. 1 cfs and 0.65 fps
c. 0.5 cfs and 1.3 fps
d. 0.5 cfs and 0.65 fps

2.30 Given the hydrograph in Exhibit 2.30a, the total volume of storm runoff is most nearly:
a. 2×10^6 ft^3
b. 4×10^6 ft^3
c. 6×10^6 ft^3
d. The total volume cannot be estimated from the information given.

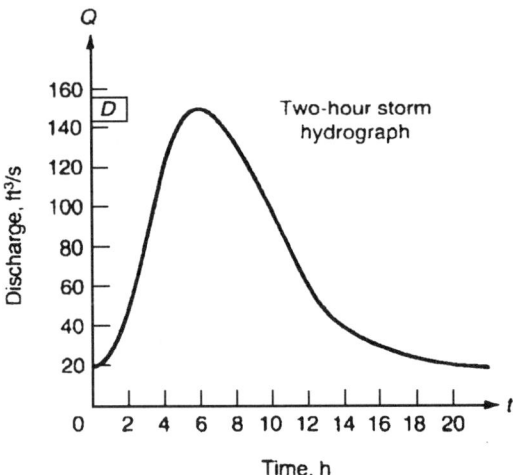

Exhibit 2.30a

Problems 2.31–2.32

Consider the schematic (*not to scale*) of a water distribution system shown in Exhibit 2.31. All nodes are at equal elevation, and all pipes are the same size, material, and age. The arrows denote the known direction of flow in certain pipes. Known head losses are also reported.

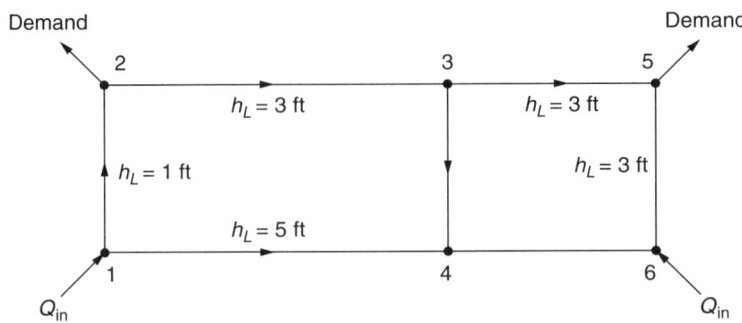

Exhibit 2.31

2.31 The head loss in pipe 3-4 is most nearly:
a. 1 ft
b. 2 ft
c. 3 ft
d. 4 ft

2.32 The flow in pipe 3-5 is _____ the flow in pipe 5-6.
   a. greater than
   b. less than
   c. equal to
   d. Cannot be determined from the information given

2.33 Which of the following are examples of disinfection by-products?
   a. PCBs
   b. THMs
   c. dioxins
   d. All of the above

2.34 A water treatment plant produces 100,000 pounds of alum residuals ("sludge") at 2 percent total solids (w/w) daily. The residuals are dewatered to 20 percent total solids (w/w) using a filter press. Given a specific gravity of sludge solids of 2.0, the volume of sludge sent to disposal daily is most nearly:
   a. 800 gallons
   b. 1000 gallons
   c. 2000 gallons
   d. 10,000 gallons

2.35 The decay of pathogens due to chlorination follows first-order kinetics. Given a flow rate of 1 MGD and a decay coefficient of 0.2 min^{-1}, the volume of a batch reactor required to obtain 99.9 percent reduction is most nearly:
   a. 15,000 gallons
   b. 25,000 gallons
   c. 35,000 gallons
   d. 45,000 gallons

2.36 Commercial land development typically impacts the hydrologic cycle by:
   a. decreasing interception
   b. decreasing infiltration
   c. decreasing transpiration
   d. all of the above

2.37 A well has been pumped at a steady rate for one year. The well has a 15 cm inside diameter and fully penetrates an unconfined aquifer that has a thickness of 50 m. The drawdown in each of two observation wells, located 20 m and 45 m from the well, is observed to be 8 m and 7 m, respectively. The aquifer is glacial outwash material and has an approximate hydraulic conductivity of 0.025 cm/s.

The expected discharge from the well is most nearly:
   a. 1 liter/s
   b. 10 liters/s
   c. 30 liters/s
   d. 80 liters/s

2.38 Given an overflow rate of 20 m/day and a flow rate of 0.2 m³/s, the detention time in a 2.5 m deep sedimentation basin with a length:width ratio of 2:1 is most nearly:
a. 40 minutes
b. 100 minutes
c. 180 minutes
d. 220 minutes

2.39 Consider a two-hydrant test performed on hydrants spaced 800 feet apart. When the flow is known to be 1.57 cfs in the 6-inch water main connecting the two hydrants, the pressure in one hydrant reads 50 psig, and the pressure in the second is 60 psig. The hydrants are at the same elevation. The pipe material is most likely:
a. PVC
b. ductile iron
c. galvanized iron
d. cast iron

2.40 Catchments X and Y drain to stormwater Inlets X and Y, respectively (Exhibit 2.40a). The 18-inch concrete pipe connecting Inlets X and Y is 1000 feet long and is laid at a slope of 2 percent. For the given Intensity-Duration curve for a ten-year return period (Exhibit 2.40b), the design intensity to be used in sizing the pipe draining the combined flow of both catchments is most nearly:
a. 1 in./hr
b. 2 in./hr
c. 4 in./hr
d. 6 in./hr

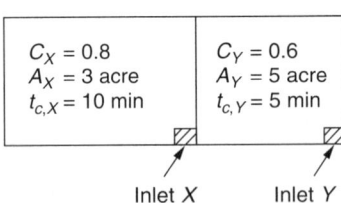

Exhibit 2.40a

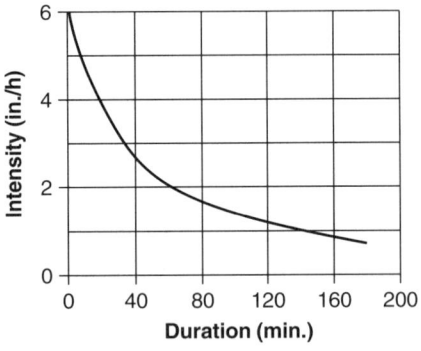

Exhibit 2.40b

CHAPTER 3

Afternoon Exam—Geotechnical Engineering

3.1 A moist soil specimen has a volume of 0.15 ft³ and weighs 18 lb. Its moisture content is 12 percent, and the specific gravity of the soil solids is 2.69.

The degree of saturation of the soil is most nearly:
a. 42%
b. 56%
c. 37%
d. 66%

3.2 The in situ moisture content of a soil is 16 percent and the moist unit weight is 102 lb/ft³. The specific gravity of soil solids is 2.75. This soil is to be excavated and transported to a construction site for use in a compacted fill. The specifications call for the soil to be compacted to a minimum dry unit weight of 103.5 lb/ft³ at the same moisture content of 16 percent. A total of 8000 cubic yards of compacted fill is needed.

The volume of soil needed to be excavated is most nearly:
a. 8200 yd³
b. 9400 yd³
c. 10,500 yd³
d. 11,200 yd³

3.3 A 10-meter-thick layer of stiff saturated clay is to be underlain by a layer of sand (Exhibit 3.3). The sand is under artesian pressure.

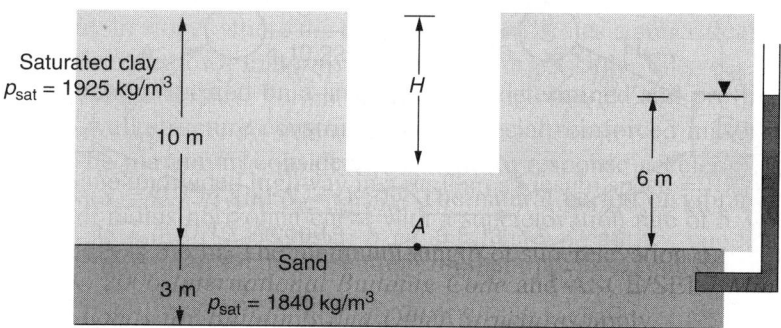

Exhibit 3.3

The maximum depth of cut H that can be made in the clay is about:
a. 4 m
b. 5 m
c. 7 m
d. 9 m

3.4 During a laboratory consolidation test, a normally consolidated clay has the following values:

Void Ratio, e	Effective Pressure, p
0.6	700 kN/m^2
0.72	350 kN/m^2

The coefficient of volume compressibility, m_v, of the soil in that pressure range is about:
a. 1.1×10^{-4} m^2/kN
b. 2.07×10^{-4} m^2/kN
c. 2.82×10^{-4} m^2/kN
d. 3.48×10^{-4} m^2/kN

3.5 A square foundation measures 2 × 2 meters in plan. The foundation is subjected to a uniformly distributed load of 200 kN/m^2.

The net stress increase in the soil due to the distributed load at a depth of 4 m below the center of the foundation (Boussinesq's theory) is nearly equal to:
a. 10 kN/m^2
b. 15 kN/m^2
c. 20 kN/m^2
d. 25 kN/m^2

3.6 A submerged, normally consolidated clay layer is 3 meters thick. The average vertical pressure on the clay layer is 90.59 kN/m^2. Due to construction of a foundation, the average vertical pressure in the clay layer increased by 27 kN/m^2. The void ratio and the compression index of the clay are 0.9 and 0.3, respectively.

The consolidation settlement due to the foundation construction is nearly equal to:
a. 40 mm
b. 47 mm
c. 54 mm
d. 61 mm

3.7 A consolidated-undrained triaxial test was performed on a normally consolidated silty clay. Following are the results:

Chamber confining pressure, $\sigma_3 = 10$ lb/in.²
Added axial (deviator) stress, $\Delta\sigma_d$, for failure of the specimen = 8 lb/in.²
Pore water pressure at failure = 6 lb/in.²

The consolidated-drained friction angle, ϕ, is about:
a. 20°
b. 25°
c. 30°
d. 35°

3.8 A frictionless retaining wall is 6 meters high and has a horizontal backfill of soft saturated cohesive soil. For the soil, given

Unit weight, $\gamma = 19$ kN/m³
Undrained cohesion, c_u (based on $\phi = 0$ condition) = 20 kN/m²

The Rankine active force per unit length of the wall after the occurrence of the tensile crack is approximately:
a. 100 kN/m
b. 150 kN/m
c. 200 kN/m
d. 250 kN/m

3.9 The inside and outside diameters of a standard spoon sampler used for field exploration are 34.93 mm and 50.8 mm, respectively.

The area ratio of the sampler is most nearly equal to:
a. 60%
b. 90%
c. 110%
d. 130%

3.10 A retaining wall with a horizontal granular backfill is shown in Exhibit 3.10a. Given: height of retaining wall, $H = 4$ m; soil friction angle, $\phi = 30°$; unit weight of soil above the groundwater table, $\gamma = 15.7$ kN/m³; and saturated unit weight of soil below the groundwater table, $\gamma_{sat} = 18.2$ kN/m³.

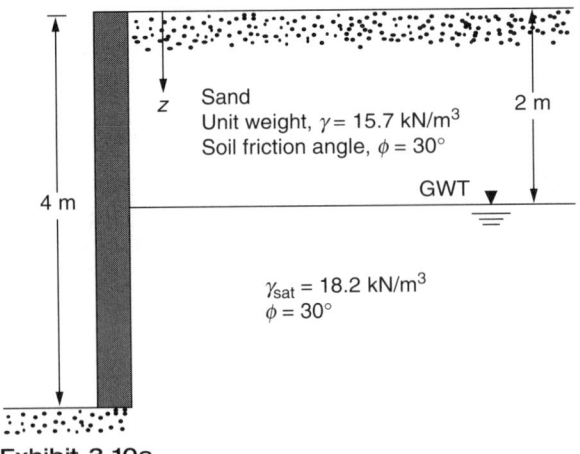

Exhibit 3.10a

If the wall is considered frictionless, the active force per unit length of the wall will be most nearly equal to:
a. 31 kN/m
b. 41 kN/m
c. 56 kN/m
d. 80 kN/m

3.11 A concrete cantilever retaining wall is shown in Exhibit 3.11. Given: unit weight of concrete, $\gamma_c = 150$ lb/ft^3. For all calculations, neglect the weight of the soil on the toe and passive pressure in the toe side. Assume Rankine active pressure along the plane ab.

The factor of safety against overturning is most nearly:
a. 2.5
b. 3.0
c. 3.5
d. 4.0

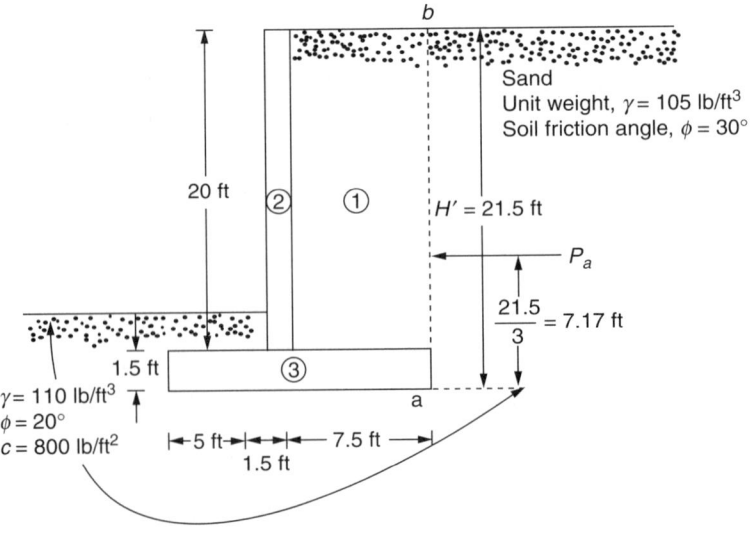

Exhibit 3.11

3.12 A concrete pile 18 × 18 in. in cross section and 100 feet long is driven into a saturated clay layer. For the clay, undrained cohesion, c_u, is 1000 lb/ft². The adhesion along the pile-clay interface $c_a = \alpha c_u = 0.75 c_u$.

The ultimate load-carrying capacity (point bearing and side resistance) is most nearly equal to:
a. 250,000 lb
b. 370,000 lb
c. 470,000 lb
d. 570,000 lb

3.13 In a field exploration program, the coring of rock was done. In one run:

Length of rock cored = 3 ft
Length of rock core excavated = 2.1 ft
Length of recovered pieces having a length of 4 in. or longer = 1.7 ft

The rock quality designator (RQD) is most nearly equal to:
a. 40%
b. 50%
c. 60%
d. 70%

3.14 A soil has a void ratio (e) of 0.71, and the specific gravity of soil solids is 2.68. The saturated unit weight of the soil will be most nearly equal to:
a. 110 lb/ft³
b. 120 lb/ft³
c. 130 lb/ft³
d. 140 lb/ft³

3.15 In a sand deposit, due to the upward flow of water, quicksand conditions occurred. Given: dry unit weight of sand = 105 lb/ft³; specific gravity of soil solids, G_s = 2.68. The critical hydraulic gradient causing quicksand conditions is most nearly equal to:
a. 1.1
b. 2.1
c. 2.3
d. 4.2

Problems 3.16–3.18
Use the flow net shown in Exhibit 3.16 for the following three problems.

3.16 The seepage through the flow channel marked II is approximately:
a. 3.1×10^{-4} m³/s/m
b. 4.6×10^{-4} m³/s/m
c. 5.2×10^{-4} m³/s/m
d. 6.4×10^{-4} m³/s/m

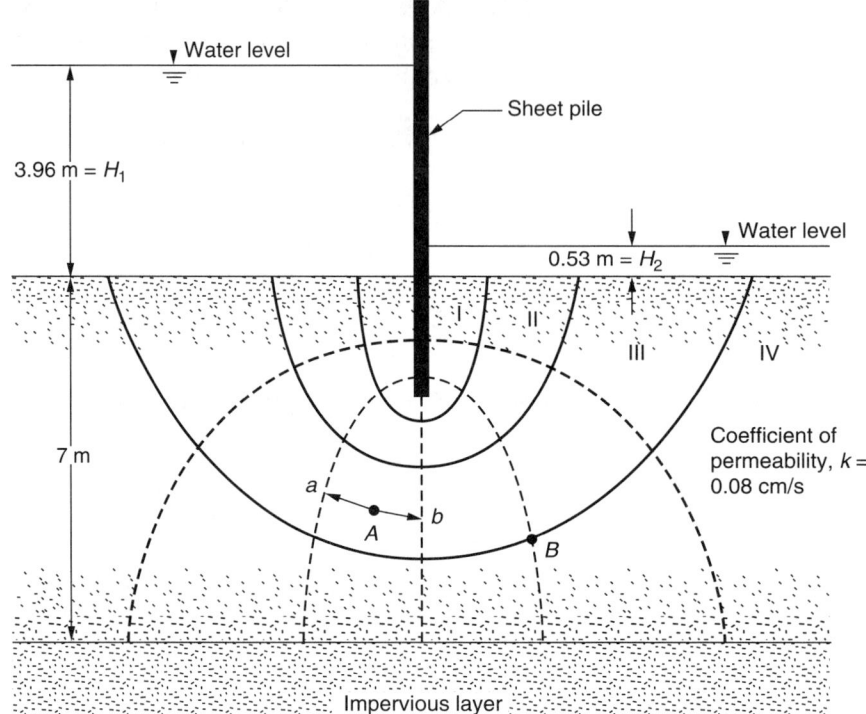

Exhibit 3.16

3.17 The total rate of seepage around the sheet pile is most nearly equal to:
a. 1×10^{-3} m³/s/m
b. 1.4×10^{-3} m³/s/m
c. 1.8×10^{-3} m³/s/m
d. 2.2×10^{-3} m³/s/m

3.18 The hydraulic gradient at A is nearly equal to:
a. 0.15
b. 0.29
c. 0.37
d. 0.48

3.19 A drained triaxial test was conducted on a sand specimen. Given: hydrostatic confining pressure, $\sigma_3 = 70$ kN/m²; axial stress, Δ_d, added to cause failure = 120 kN/m². The drained friction angle, ϕ, of the sand is nearly equal to:
a. 28°
b. 32°
c. 36°
d. 40°

3.20 A Shelby tube having internal (D_i) and external (D_o) diameters of 1.875 in. and 2 in., respectively, was used in a field exploration to obtain a clay soil sample. The area ratio of the Shelby tube is most nearly equal to:
a. 10%
b. 12%
c. 14%
d. 16%

3.21 Consider a frictionless retaining wall 10 ft high with a horizontal backfill of sand. The unit weight of the sand is 110 lb/ft^3, and it has an angle of friction $\phi = 36°$. The intensity of Rankine active pressure at the bottom of the wall is nearly equal to:
   a. 150 lb/ft^2
   b. 290 lb/ft^2
   c. 450 lb/ft^2
   d. 600 lb/ft^2

3.22 Refer to the retaining wall described in Problem 3.21. The active force per unit length of the wall is nearly equal to:
   a. 1450 lb/ft
   b. 1950 lb/ft
   c. 2450 lb/ft
   d. 2950 lb/ft

3.23 Refer to the retaining wall described in Problem 3.21. The Rankine passive force per unit length of the wall is nearly equal to:
   a. 10 kips/ft
   b. 15 kips/ft
   c. 21 kips/ft
   d. 25 kips/ft

3.24 A normally consolidated clay layer is 10 ft thick and is located below the ground water table. Given: average effective stress on the clay layer, p_o, is 1600 lb/ft^2; in situ void ratio, $e_0 = 1.1$; compression index, $C_c = 0.27$; swell index, $C_s = 0.05$. If the average effective stress on the clay layer is increased by 1000 lb/ft^2, the primary consolidation settlement, S, will be approximately:
   a. 1 in.
   b. 2.2 in.
   c. 3.25 in.
   d. 4.5 in.

3.25 The laboratory consolidation test on a clay soil yielded the following:

Effective Pressure p, lb/ft	Void Ratio
2000	1.1
4000	0.98

The compression index, C_c, of the soil is about:
   a. 0.25
   b. 0.31
   c. 0.4
   d. 0.45

3.26 A braced cut is shown in Exhibit 3.26a. Assume that the sheet piles are hinged at the strut level B. The spacing of struts is 3 m center-to-center. Assume Peck's earth pressure envelope (1969); that is, $\sigma_a = 0.65\gamma H K_a$ (K_a = Rankine active earth pressure coeffficient). The strut load at level A is nearly equal to:
a. 250 kN
b. 390 kN
c. 450 kN
d. 500 kN

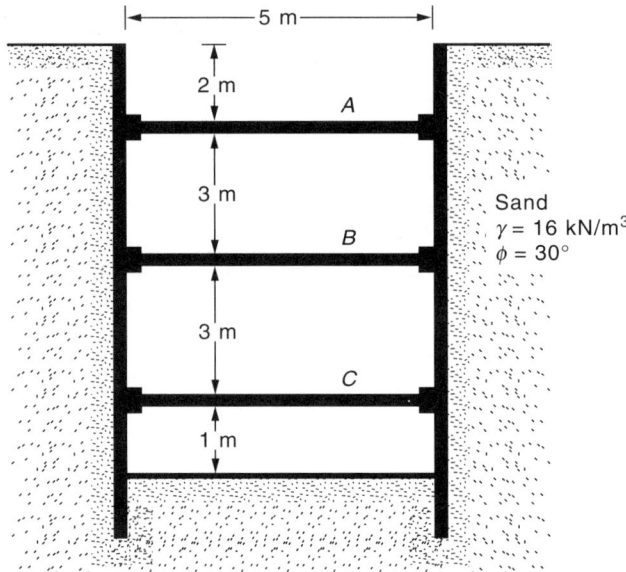

Exhibit 3.26a

3.27 Exhibit 3.27 shows the loading acting on one line of a symmetrical group of piles supporting a structure. The vertical load indicated includes the self-weight of the rigid pile cap. The maximum axial force produced in a pile is most nearly:
a. 22 kips
b. 19 kips
c. 16 kips
d. 13 kips

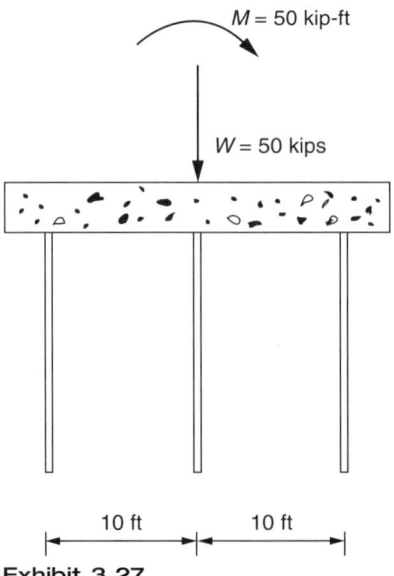

Exhibit 3.27

3.28 A nominal 16-inch-square, solid-grouted, concrete block masonry column has an effective height of 40 feet. The column is reinforced with four No. 7, Grade 60 reinforcing bars.

Codes:
ACI 530-05, *Building Code Requirements for Masonry Structures*

Materials:
Masonry strength, $f'_m = 1500$ psi
Modulus of elasticity = 1000 ksi
Reinforcing steel = Grade 60, ASTM A-615

The maximum axial load, including its own weight, that the column can support using Allowable Stress Design is most nearly:
a. 52 kips
b. 54 kips
c. 56 kips
d. 58 kips

3.29 The reinforced concrete footing shown in Exhibit 3.29 has an effective depth of 15 inches and supports a column with a factored load of 250 kips.

Design Data:
Normal weight concrete, $f'_c = 3000$ psi
Reinforcing steel = Grade 60, ASTM A-615
Weight of concrete, $\gamma_c = 150$ pcf

The strength design provisions of the ACI *Building Code Requirements for Structural Concrete*, ACI 318-05, apply.

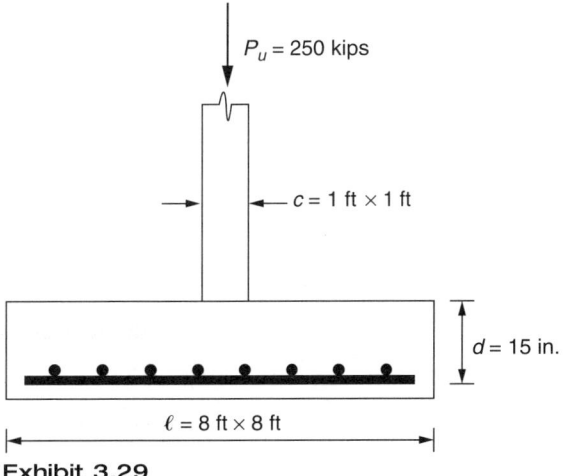

Exhibit 3.29

The capacity of the footing to resist punching shear is most nearly:
a. 265 kips
b. 280 kips
c. 300 kips
d. 320 kips

3.30 A nominal 16-inch-square, solid-grouted, concrete block masonry column has an effective height of 40 feet. The column is reinforced with four No. 7, Grade 60 reinforcing bars.

Codes:
ACI 530-05, *Building Code Requirements for Masonry Structures*

Materials:
Masonry strength, $f'_m = 1500$ psi
Modulus of elasticity = 1,000 ksi
Reinforcing steel = Grade 60, ASTM A-615

The maximum allowable spacing of No. 3 lateral ties is most nearly:
a. 8 inches
b. 14 inches
c. 16 inches
d. 18 inches

3.31 The reinforced concrete footing shown in Exhibit 3.31 has an effective depth of 16 inches and supports a column with a factored load of 300 kips.

Design Data:
Normal weight concrete, $f'_c = 3000$ psi
Reinforcing steel = Grade 60, ASTM A-615
Weight of concrete, $\gamma_c = 150$ pcf

The strength design provisions of the ACI *Building Code Requirements for Structural Concrete*, ACI 318-05, apply.

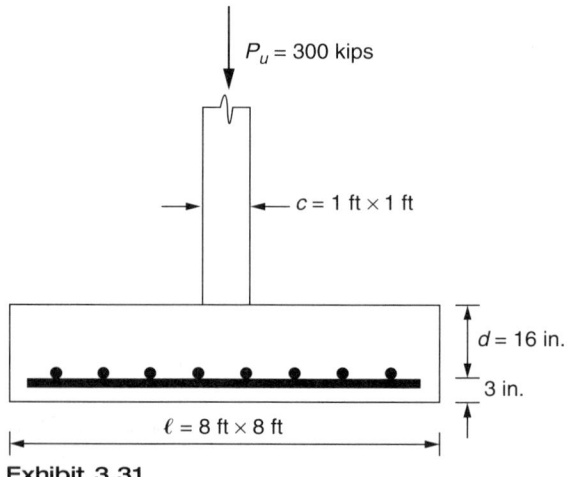

Exhibit 3.31

The area of tension reinforcement required in the footing is most nearly:
a. 3.26 in.^{2}
b. 3.27 in.^{2}
c. 3.28 in.^{2}
d. 3.29 in.^{2}

3.32 A 4 × 12 Douglas Fir–Larch ledger is bolted to a concrete wall with 7/8-inch-diameter bolts at 4-foot centers.

The ICC *2006 International Building Code* and *2005 NDS for Wood Construction* apply.

Design data:
Moisture content is less than 19 percent.
The capacity of the bolt in the concrete wall does not govern.

Based on the 7/8-inch-diameter bolt design capacity in the ledger, the maximum vertical load that the ledger can support is most nearly:
a. 300 pounds/foot
b. 270 pounds/foot
c. 240 pounds/foot
d. 210 pounds/foot

3.33 The reinforced concrete cantilever retaining wall shown in Exhibit 3.33 retains fill with an equivalent fluid pressure of 30 pounds per square foot per foot.

Design Data:
Normal weight concrete, $f'_c = 3000$ psi
Reinforcing steel = Grade 60, ASTM A-615
Weight of concrete, $\gamma_c = 150$ pcf

The strength design provisions of the ACI *Building Code Requirements for Structural Concrete*, ACI 318-05, apply.

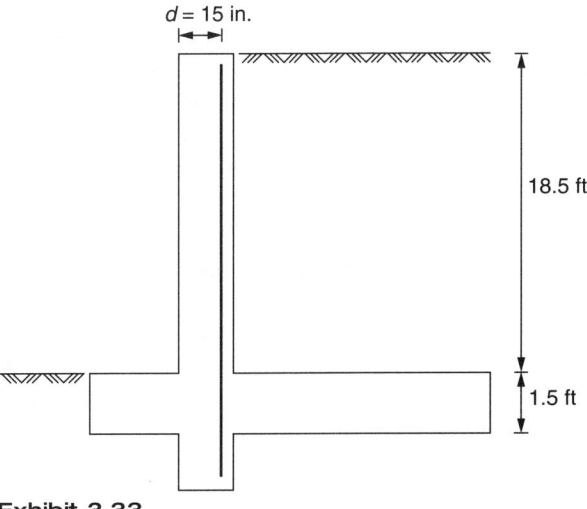

Exhibit 3.33

The area of tension reinforcement required in the stem wall is most nearly:
a. 0.78 in.²/ft
b. 0.79 in.²/ft
c. 0.80 in.²/ft
d. 0.81 in.²/ft

3.34 The reinforced concrete cantilever retaining wall shown in Exhibit 3.34 retains fill with an equivalent fluid pressure of 30 pounds per square foot per foot. The passive earth pressure is equivalent to a fluid pressure of 300 pounds per square foot per foot, and the coefficient of friction on the underside of the base is 0.4. The total weight of the retaining wall plus backfill is 20 kips.

Design Data:
Normal weight concrete, f'_c = 3000 psi
Reinforcing steel = Grade 60, ASTM A-615
Weight of concrete, γ_c = 150 pcf

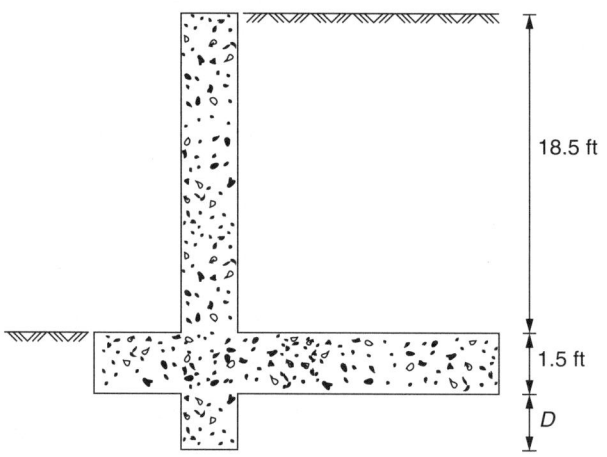

Exhibit 3.34

The depth of shear key required to provide a factor of safety of 1.5 in the retaining wall is most nearly:
a. 1.7 feet
b. 1.5 feet
c. 1.3 feet
d. 1.1 feet

3.35 The table below gives earthwork cross section areas for three stations.

Station	Cut End Area, ft²	Fill End Area, ft²
50 + 00	120.25	0
50 + 40	60.75	0
51 + 20	15.87	25.5

If the average end area method is used to compute the earthwork volume, the total volume of excavated material is approximately:
a. 197 yd³
b. 228 yd³
c. 248 yd³
d. 685 yd³

3.36 Exhibit 3.36 shows the coordinates (in meters) of points on an earthwork cross section, with the roadway elevation at the centerline taken as the origin.

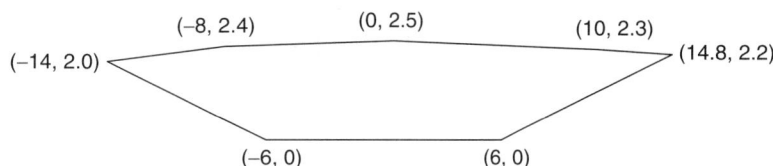

Exhibit 3.36

The area of the cross section is approximately:
a. 35 m^2
b. 40 m^2
c. 45 m^2
d. 50 m^2

Problems 3.37–3.40
The release of a toxic organic compound from an underground storage tank has just been identified. Initial tests have determined the following:

groundwater table = 12 feet bgs (below ground surface)
confining layer = 40 feet bgs
width of contaminated area = 100 feet
length of contaminated area = 600 feet
average contaminant concentration in the aqueous phase (C_{aq}) = 2.3 mg/L
soil water partition coefficient (K_{SW}) = 0.4
soil porosity = 32%
contaminant specific gravity = 0.87

3.37 Observation wells at radii of 30 feet and 40 feet have drawdowns of 4.6 feet and 0.3 foot, respectively, when water is pumped from the extraction well at a rate of 5 gpm. Estimate the soil hydraulic conductivity.
a. 146.4 ft/yr
b. 133.6 ft/yr
c. 119.3 ft/yr
d. 101.8 ft/yr

3.38 Estimate the total volume of contaminant released, assuming contamination is only in the saturated zone in the unconfined aquifer. You may also assume a rectangular contaminated zone with the dimensions given above.
a. 4.7 gallons
b. 35 gallons
c. 115 gallons
d. 254 gallons

3.39 Estimate the fraction of organic matter in the soil if lab experiments find the contaminant concentrations in an octanol/water system are $C_o = 123$ mg/L and $C_w = 4.6$ mg/L. You may assume the following relationship applies: $\log(K_{oc}) = \log(K_{ow}) - 0.21$
  a. 0.92%
  b. 1.51%
  c. 1.85%
  d. 2.43%

3.40 Estimate the time required for the contaminant to travel to the nearest drinking well (located 300 yards away). You may assume one-dimensional flow, single-source contamination with no biological removal mechanisms and negligible dispersion effects.
  a. 3.7 years
  b. 6.1 years
  c. 14.4 years
  d. 22.7 years

CHAPTER 4

Afternoon Exam—Structural Engineering

4.1 The regular prestressed concrete bridge shown in Exhibit 4.1 is continuous over three spans of 80 feet and is situated on a strategic route. AASHTO, *Standard Specifications for Highway Bridges,* 17th ed. (2002), applies.

The design loading is HS20-44 truck loading.

The soil profile at the site consists of a 220-foot depth of stiff soil. The acceleration coefficient for the site is $A = 0.5$.

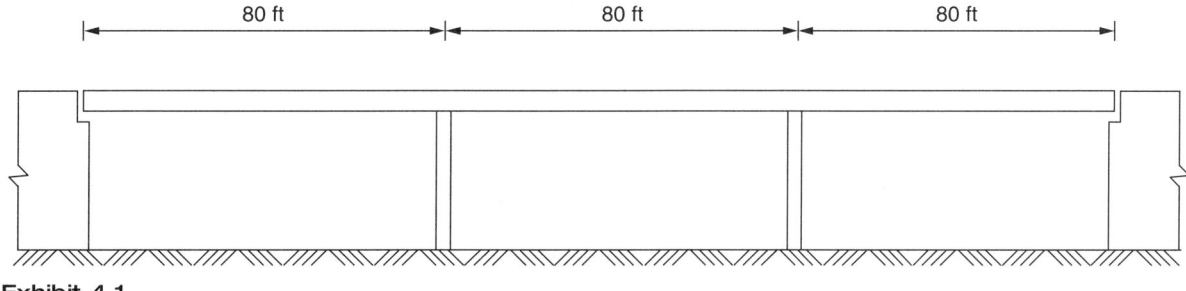

Exhibit 4.1

The required seismic analysis procedure for the bridge is:
a. None required
b. Procedure 1
c. Procedure 1 or 2
d. Procedure 3

4.2 Exhibit 4.2 shows the service loads acting on a fixed-ended beam. The distributed load indicated is dead load and includes the beam self-weight. The concentrated load at midspan is floor live load. Continuous lateral support is provided to the beam.

Design data:
Steel members to ASTM A992, $F_y = 50{,}000$ psi

AISC *Steel Construction Manual,* 13th ed. (2006), applies.

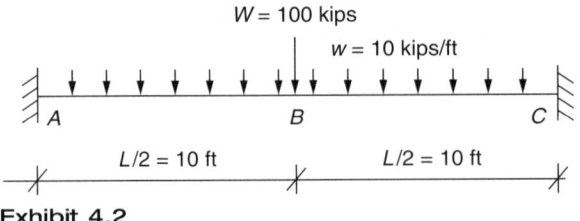

Exhibit 4.2

You may select **either** the ASD **or** the LRFD option.

ASD Option

Using plastic design, the lightest W30 section beam (plf) that can support the applied loads is most nearly:
a. 90
b. 99
c. 108
d. 116

LRFD Option

The lightest W30 section beam (plf) that can support the applied loads is most nearly:
a. 90
b. 99
c. 108
d. 116

4.3 A wood-framed workshop is shown in Exhibit 4.3.

The ICC *2006 International Building Code* and the following data apply:

Roof framing members = 2 × Douglas Fir–Larch select structural
Moisture content < 19%
Plywood roof diaphragm = $^{15}/_{32}$ inch Structural I

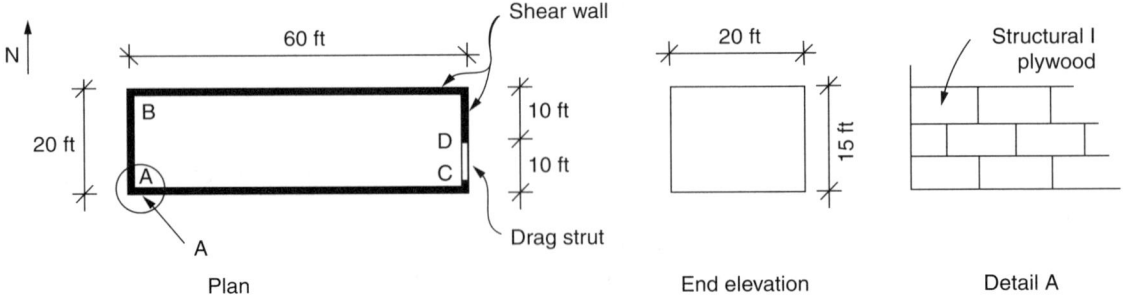

Exhibit 4.3

The diaphragm is blocked at plywood panel edges, and seismic effects govern. The service-level seismic design load in the north-south direction at the roof diaphragm level is 280 plf. The structure is assigned to seismic design category C.

Using a 2 × drag strut in a single shear connection, the number of 16d common wire nails required to connect the drag strut to the shear wall is most nearly:
a. 13
b. 15
c. 17
d. 19

4.4 The interior reinforced concrete floor beam shown in Exhibit 4.4 is continuous over three clear spans of 35 feet. The ends of the beam are integral with the outer columns. All columns are 12 inches wide.

Design data:
Normal weight concrete, $f'_c = 3000$ psi
Reinforcing steel = Grade 60, ASTM A-615
Beam width = 12 in.
Effective depth = 21.5 in.
Dead load (not including beam weight) = 1.0 klf
Live load = 1.0 klf
Weight of concrete, $\gamma_c = 150$ pcf

The strength design provisions of the ACI *Building Code Requirements for Structural Concrete*, ACI 318-05, apply.

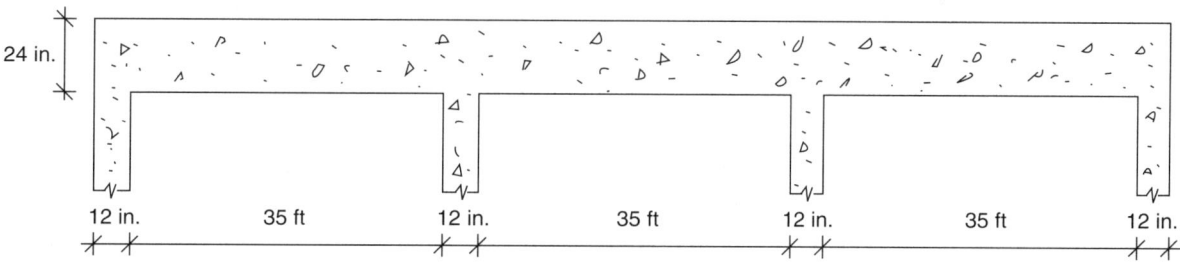

Exhibit 4.4

Using ACI 318 approximate moment coefficients, the factored positive moment, in kip-feet, in the end span is most nearly:
a. 280
b. 300
c. 320
d. 340

4.5 A fire station, located on a site with an undetermined soil profile, is a bearing wall structure constructed with special reinforced masonry shear walls. The maximum considered earthquake response acceleration parameters are $S_s = 0.75g$ and $S_1 = 0.50g$. The natural period of vibration of the building is $T_a = 0.15$ second.

The ICC *2006 International Building Code* and ASCE/SEI 7 *Minimum Design Loads for Buildings and Other Structures* apply.

The seismic design category for the building is most nearly:
a. B
b. C
c. D
d. E

4.6 Exhibit 4.6 shows the interior girder of a reinforced concrete bridge that is simply supported over an effective span of 40 feet. AASHTO *Standard Specifications for Highway Bridges,* 17th ed. (2002), applies.

Design data:
Normal weight concrete, f'_c = 4000 psi
Reinforcement bars, f_y = 60,000 psi
Beam width, b_w = 12 in.
Effective depth, d = 40 in.

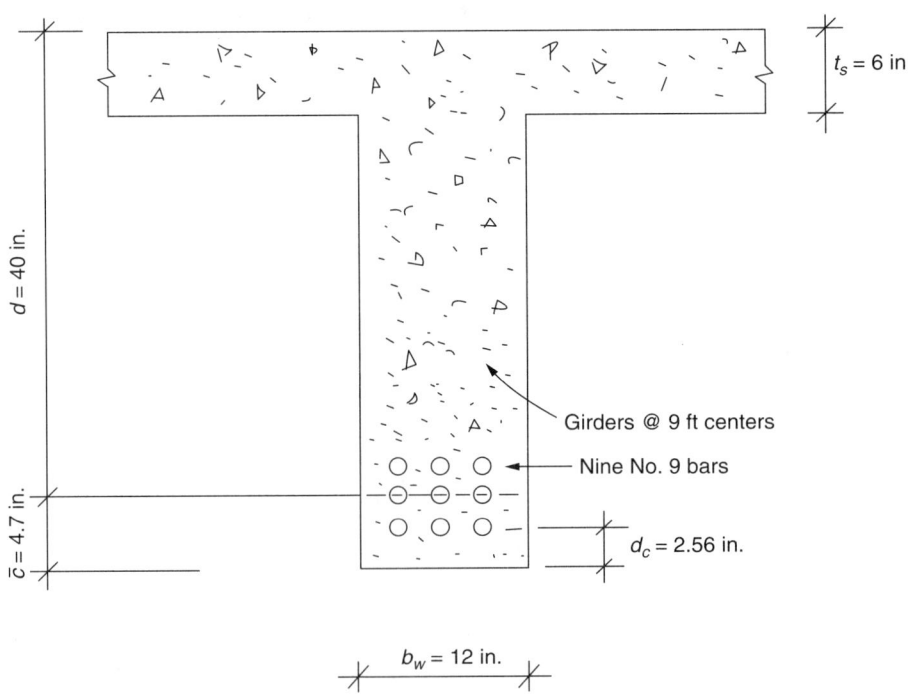

Exhibit 4.6

The design loading is HS20-44 truck loading. The live load service moment on the beam, including impact, is 439 kip-feet. The dead load service moment on the beam is 350 kip-feet.

Assuming that the lever arm of the elastic resisting moment is the distance from centroid of reinforcement to center of slab, the maximum allowable fatigue stress range, in ksi, in the reinforcement is most nearly:
a. 13
b. 15
c. 17
d. 19

4.7 The prestressed concrete beam shown in Exhibit 4.7 supports a uniformly distributed load, including the weight of the beam, of 2 kips per foot. The prestressing tendon has a parabolic profile with zero eccentricity at the ends and a drape of 10 inches.

Design data:
Normal weight concrete, $f'_c = 6000$ psi
Beam width = 12 in.
Beam depth = 30 in.
Weight of concrete, $\gamma_c = 150$ pcf

The design provisions of the ACI *Building Code Requirements for Structural Concrete,* ACI 318-05, apply.

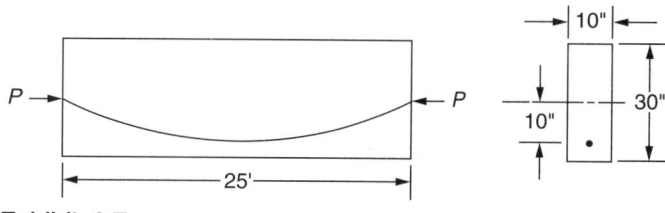

Exhibit 4.7

The magnitude of the prestressing force P, in kips, required to balance exactly the applied load on the beam is most nearly:
a. 170
b. 180
c. 190
d. 200

4.8 The single-plate shear connector, shown in Exhibit 4.8, consists of a $3\frac{1}{2} \times \frac{1}{2} \times 25$ inch plate. Eight $\frac{7}{8}$-inch-diameter bolts are provided with standard holes in a bearing-type connection with threads excluded from the shear plane. The capacity of the connection is governed by the strength of the $\frac{3}{16}$-inch fillet weld.

Design data:
Steel members to ASTM A992, $F_y = 50{,}000$ psi
Bolts = A490X, 3 in. spacing, $1\frac{3}{4}$ in. edge distance
Welds = E70XX, $\frac{3}{16}$ in. fillet weld, shielded metal arc
Weld design = elastic vector method

AISC *Steel Construction Manual,* 13th ed. (2006), applies.

You may select **either** the ASD **or** the LRFD option.

ASD Option

The maximum load, in kips, that the connection can support is most nearly:
a. 100
b. 110
c. 120
d. 130

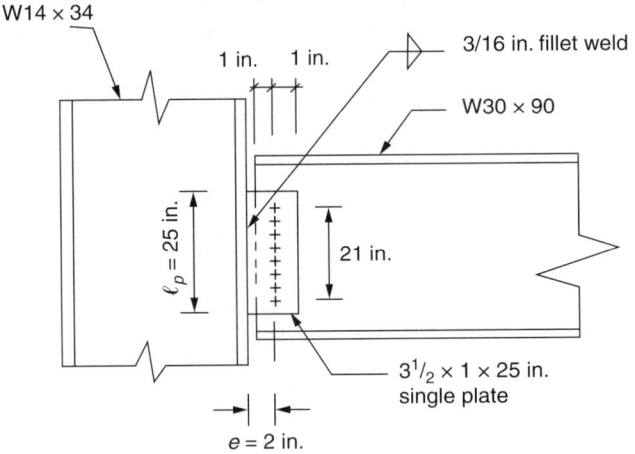

Exhibit 4.8

LRFD Option

The design strength, in kips, of the connection is most nearly:
a. 160
b. 170
c. 180
d. 190

4.9 A solid, normal-weight, reinforced concrete slab for an assembly area with movable seats spans between 8-inch-wide beams at 15-foot centers over five bays. The discontinuous ends of the slab are unrestrained. The superimposed dead load, due to finishes and services, is 40 pounds per square foot. The concrete strength is 3000 pounds per square inch, and all reinforcement consists of Grade 60 No. 4 bars. The slab does not support any deflection-sensitive partitions or ceilings.

Design data:
Normal weight concrete, $f'_c = 3000$ psi
Reinforcing steel = Grade 60, ASTM A-615
Slab depth, $h = 7.5$ in.
Weight of concrete, $\gamma_c = 150$ pcf

The strength design provisions of the ACI *Building Code Requirements for Structural Concrete,* ACI 318-05; the ASCE/SEI 7 *Minimum Design Loads for Buildings and Other Structures;* and the ICC *2006 International Building Code* apply.

Using ACI 318 approximate moment coefficients, the factored moment, in kip-feet, at the first interior support is most nearly:
a. 6.7
b. 7.0
c. 7.3
d. 7.6

4.10 A water tank mounted on a braced tower is shown in Exhibit 4.10. Wind pressure on the braced tower and the concrete footings may be neglected.

The ICC *2006 International Building Code* and the following data apply:

Weight of tower + empty tank, W_e = 24 kips
Weight of tower + full tank, W_f = 224 kips
Weight of concrete, γ_c = 150 pcf
Wind pressure on the tank, p = 34.59 psf

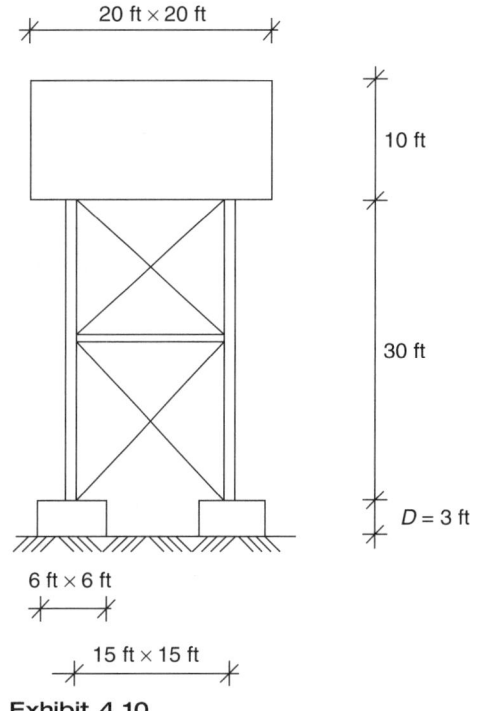

Exhibit 4.10

The maximum uplift force in kips on one leg of the tower is most nearly:
a. 7
b. 8
c. 9
d. 10

4.11 Exhibit 4.11 shows the interior girders of an 80-foot-span, simply supported composite bridge that is located on a major highway with an average daily truck traffic of 2700. The concrete flange has a 28-day compressive strength of 3000 pounds per square inch, and the steel girder is Grade A36 steel. The bridge carries two lanes of traffic, and the steel girders were fully propped before casting the concrete flange.

AASHTO *Standard Specifications for Highway Bridges,* 17th ed. (2002), applies.

The design loading is HS20-44 truck loading.
Shear connectors are $7/8$-inch-diameter welded steel studs.
The ultimate strength of one $7/8$-inch-diameter welded steel stud, S_u, is 30.6 kips.

Fatigue effects need not be considered.

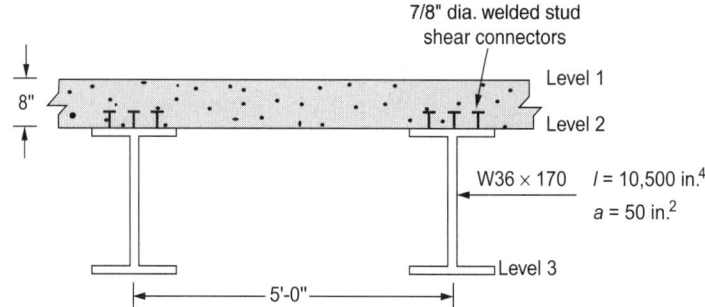

Exhibit 4.11

The total number of $7/8$-inch-diameter welded studs required between midspan and the support is most nearly:
a. 39
b. 41
c. 44
d. 47

4.12 A 12-inch-square column is supported by a 3-foot-wide rectangular base as shown in Exhibit 4.12. The column applies to the base an axial service load of 80 kips and a service bending moment of 29 kip-feet parallel to the longitudinal axis of the base.

Design data:
Normal weight concrete, $f'_c = 3000$ psi
Reinforcing steel = Grade 60, ASTM A-615
Weight of concrete, $\gamma_c = 150$ pcf

The strength design provisions of the ACI *Building Code Requirements for Structural Concrete,* ACI 318-05, apply.

Neglecting the self-weight of the base, the required base length, in feet, for an allowable soil bearing pressure of 5000 psf is most nearly:
a. 5
b. 6
c. 7
d. 8

4.13 A nominal 8-inch-wide, solid-grouted, concrete block masonry beam has an effective depth, d, of 37 inches. The maximum shear force at the critical section for shear is $V = 14.5$ kips.

ACI 530-05, *Building Code Requirements for Masonry Structures,* applies.

Design data:
Masonry strength, $f'_m = 1500$ psi
Modulus of elasticity = 1000/ksi
Reinforcing steel = Grade 60, ASTM A-615
Shear reinforcement = No. 4 bars (single arm)

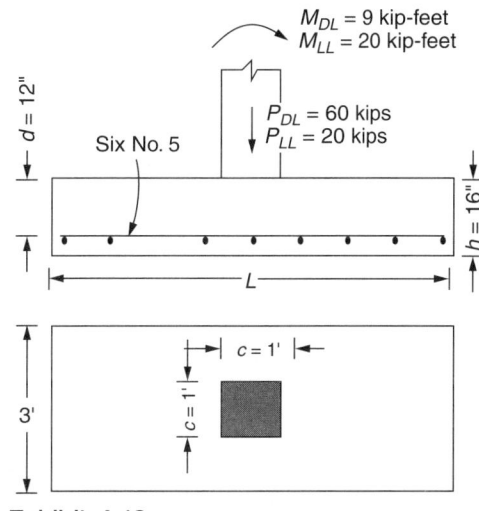

Exhibit 4.12

The required spacing of shear reinforcement at the critical section using Allowable Stress Design is most nearly:
a. 10
b. 12
c. 14
d. None required

4.14 A single-story wood-framed structure is used as an office building. The soil profile at the site consists of soft rock with a shear wave velocity of 2000 feet per second.

The ICC *2006 International Building Code* applies.

The applicable site classification is:
a. B
b. C
c. D
d. E

4.15 Exhibit 4.15 shows the loading acting on one line of a symmetrical group of piles supporting a structure. The vertical load indicated includes the self-weight of the rigid pile cap. The maximum axial force produced in a pile is most nearly:
a. 22 kips
b. 19 kips
c. 16 kips
d. 13 kips

Problems 4.16–4.20
Exhibit 4.16 shows the interior girders of a simply supported 40-foot-span steel composite bridge. AASHTO *Standard Specifications for Highway Bridges,* 17th ed. (2002), applies.

Design Data:
Normal weight concrete, $f'_c = 4000$ psi
Reinforcement bars, $f_y = 60,000$ psi
The design loading is HS20-44 truck loading.

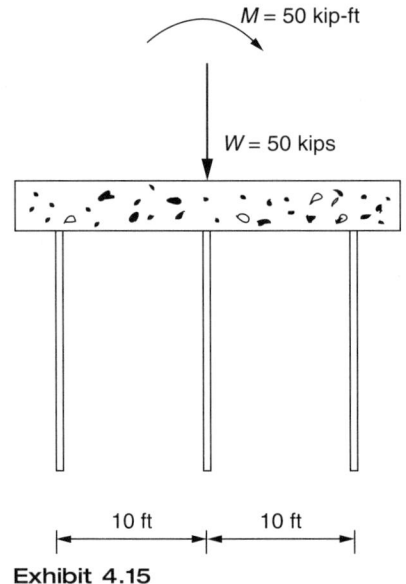

Exhibit 4.15

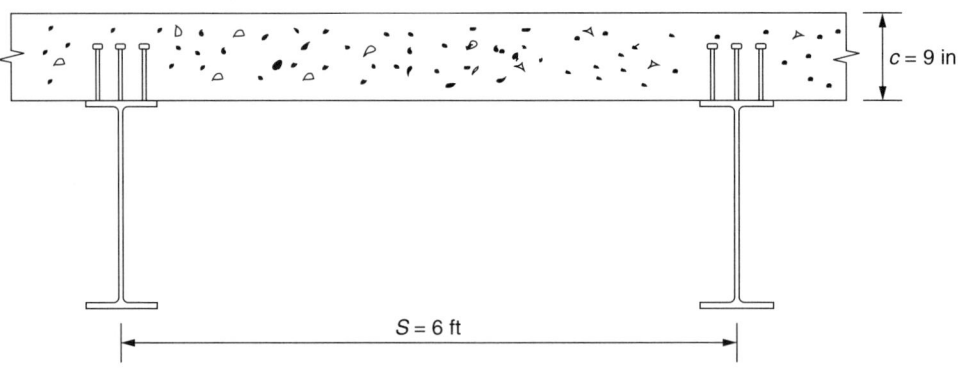

Exhibit 4.16

4.16 The effective width of the concrete slab is most nearly:
 a. 60 inches
 b. 72 inches
 c. 108 inches
 d. 120 inches

4.17 The impact fraction is most nearly:
 a. 0.28
 b. 0.29
 c. 0.30
 d. 0.31

4.18 The girder distribution factor is most nearly:
 a. 1.01
 b. 1.05
 c. 1.09
 d. 1.13

4.19 The impact fraction is $I = 0.30$, and the girder distribution factor is $G = 1.09$.

The live load plus impact moment distributed to an interior girder is most nearly:
a. 320 kip-ft
b. 325 kip-ft
c. 330 kip-ft
d. 335 kip-ft

4.20 The impact fraction is $I = 0.30$, and the girder distribution factor is $G = 1.09$.

The end reaction on an interior girder is most nearly:
a. 35 kips
b. 37 kips
c. 39 kips
d. 41 kips

4.21 A single-story wood-framed structure is used as an office building. The site classification at the location of this structure is site classification D. The maximum considered earthquake response accelerations are

$$S_S = 1.25g$$
$$S_1 = 0.50g$$

The ICC *2006 International Building Code* applies.

The applicable site coefficients for this structure are most nearly:
a. $F_a = 1.00$, $F_v = 1.00$
b. $F_a = 1.00$, $F_v = 1.30$
c. $F_a = 1.00$, $F_v = 1.50$
d. $F_a = 0.90$, $F_v = 2.40$

4.22 A single-story wood-framed structure is used as an office building. The maximum considered earthquake response accelerations at the location of the building are

$$S_S = 1.25g$$
$$S_1 = 0.50g$$

The site coefficients for this structure are

$$F_a = 1.00$$
$$F_v = 1.30$$

The ICC *2006 International Building Code* applies.

The design spectral response accelerations for the structure are most nearly:
a. $S_{DS} = 0.83$, $S_{D1} = 1.34$
b. $S_{DS} = 0.83$, $S_{D1} = 0.37$
c. $S_{DS} = 0.83$, $S_{D1} = 0.40$
d. $S_{DS} = 0.83$, $S_{D1} = 0.43$

4.23 A single-story wood-framed structure with a roof height of 15 feet is used as an office building.

The ICC *2006 International Building Code* and ASCE 7-05 apply.

The approximate fundamental period is most nearly:
a. 0.15 second
b. 0.17 second
c. 0.19 second
d. 0.21 second

4.24 Both flanges of a W14 × 68 are connected by transverse welds to gusset plates, and a tensile force is applied along the longitudinal axis of the beam.

Design Data:
Steel members to A992 Grade 50, F_y = 50,000 psi
The capacity of the transverse welds do not govern.

AISC-*Steel Construction Manual,* 13th ed. (2006), Chapter N applies.

You may select **either** the ASD **or** the LRFD option.

ASD Option

The maximum tensile force that the connection can support is most nearly:
a. 490 kips
b. 480 kips
c. 470 kips
d. 460 kips

LRFD Option

The maximum factored tensile force that the connection can support is most nearly:
a. 740 kips
b. 720 kips
c. 700 kips
d. 680 kips

4.25 A single-story wood-framed structure is used as an office building. The design spectral response accelerations for the structure are

$$S_{DS} = 0.80$$
$$S_{D1} = 0.50$$

The ICC *2006 International Building Code* and ASCE 7-05 apply.

The applicable seismic design category is most nearly:
a. B
b. C
c. D
d. E

4.26 The reinforced concrete beam shown in Exhibit 4.26 is reinforced with Grade 60 bars at the positions indicated.

Design Data:
Normal weight concrete, f'_c = 3000 psi
Reinforcing steel = Grade 60, ASTM A-615
Beam width = 15 in.
Weight of concrete, γ_c = 150 pcf

The strength design provisions of the ACI *Building Code Requirements for Structural Concrete,* ACI 318-05, apply.

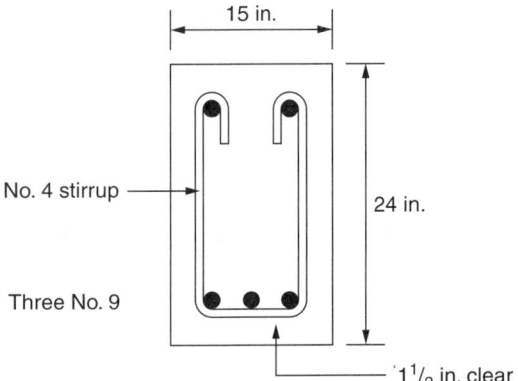

Exhibit 4.26

The maximum spacing allowed for the tension reinforcement to comply with ACI Section 10.6 is most nearly:
a. 4 inches
b. 6 inches
c. 8 inches
d. 10 inches

4.27 A soil has a porosity (n) of 0.4, moisture content (w) of 10 percent, and moist unit weight (γ) of 17.8 kN/m³. The specific gravity of the soil solids (G_s) is nearly equal to:
a. 2.55
b. 2.65
c. 2.75
d. 2.85

4.28 The maximum dry unit weight of compaction of a soil in the field should be 118 lb/ft³. The actual compacted soil has a moist unit weight (γ) of 122 lb/ft³ and a moisture content of 12 percent. The relative compaction (R) of the soil is most nearly equal to:
a. 70%
b. 80%
c. 90%
d. 98%

4.29 A soil profile in a sand deposit is shown in Exhibit 4.29. Given: specific gravity of soil solids, $G_s = 2.65$; void ratio, $e = 0.5$. The effective stress at A is nearly equal to:
a. 55 kN/m^2
b. 67 kN/m^2
c. 72 kN/m^2
d. 48 kN/m^2

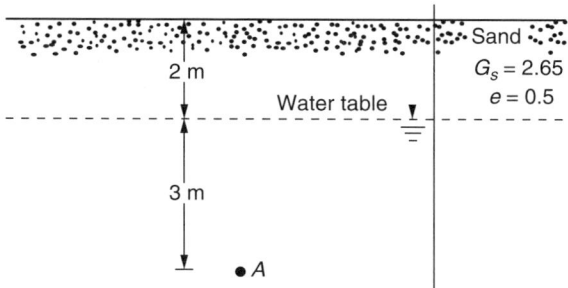

Exhibit 4.29

4.30 A dry sand specimen was tested in a direct shear equipment under a normal stress of 15 lb/in.^{2}. The specimen failed under a shear stress of 11 lb/in.^{2}. The friction angle, ϕ, of the sand is about:
a. 26°
b. 28°
c. 32°
d. 36°

4.31 Consider a frictionless retaining wall 10 ft high with a horizontal backfill of sand. The unit weight of the sand is 110 lb/ft^3, and it has an angle of friction $\phi = 36°$. The intensity of Rankine active pressure at the bottom of the wall is nearly equal to:
a. 150 lb/ft^2
b. 290 lb/ft^2
c. 450 lb/ft^2
d. 600 lb/ft^2

4.32 A frictionless retaining wall is shown in Exhibit 4.32a. The active force per unit length of the wall is nearly equal to:
a. 28 kN/m
b. 38 kN/m
c. 48 kN/m
d. 58 kN/m

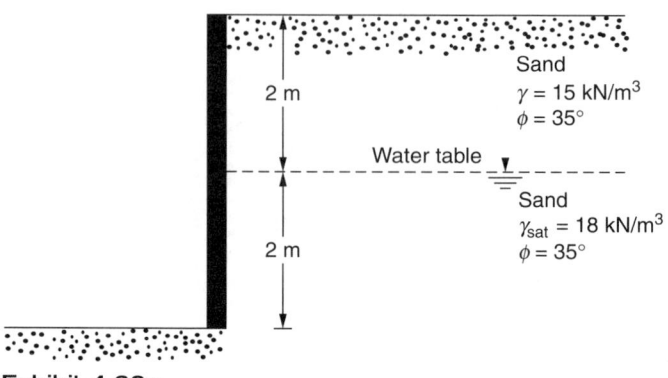

Exhibit 4.32a

4.33 A shallow square foundation has the following specifications:

Foundation: width, $B = 3$ ft
depth of foundation, $D_f = 3$ ft

Soil: unit weight, $\gamma = 115$ lb/ft^3
angle of friction, $\phi = 28°$
cohesion, $c = 400$ lb/ft^2

The ultimate load that the foundation can carry is nearly equal to (use Terzaghi's bearing capacity equation: $N_c = 31.61$, $N_q = 17.81$, $N_\gamma = 13.7$):
a. 110 kip
b. 160 kip
c. 300 kip
d. 220 kip

4.34 A continuous shallow foundation supported by a sand layer is subjected to eccentric vertical load. Given:

Foundation: width, $B = 4$ ft
depth of foundation, $d = 3$ ft
load eccentricity, $e = 0.4$ ft

Soil: unit weight, $\gamma = 120$ lb/ft^3
angle of friction, $\phi = 35°$

Use Meyerhof's effective area method and ignore the depth factor. Given: $N_q = 33.3$; $N_\gamma = 48.03$. The ultimate load per unit length that the foundation can carry is nearly:
a. 45 kip/ft
b. 55 kip/ft
c. 62 kip/ft
d. 68 kip/ft

4.35 A 30-ft-long prestressed concrete pile 18 × 18 in. in cross section is driven into a saturated clay layer. For the clay-pile interface, the adhesion factor, α, is 0.5. Given: the undrained shear strength of clay $c_u = 1600$ lb/ft^2. The ultimate load-carrying capacity of the pile is nearly equal to:
a. 90 kips
b. 120 kips
c. 150 kips
d. 175 kips

4.36 A 60-ft-long prestressed concrete pile 18 × 18 in. in cross section is driven into a sand layer. The unit weight of sand is 115 lb/ft^3. The soil friction angle is 35°. Based on Meyerhof's theory (1976), the ultimate point bearing load, Q_p, is nearly (use $N^*_q \approx 120$):
a. 150 kips
b. 190 kips
c. 230 kips
d. 330 kips

4.37 A two-lane highway has 3.6 m lanes and 2.4 m paved shoulders. The radius of curvature of the centerline of the roadway is 400 m. The line of sight is blocked by vegetation that has been trimmed to a point 2 m beyond the outer edge of the shoulder. The stopping sight distance on this curve is approximately:
a. 100 m
b. 115 m
c. 130 m
d. 145 m

4.38 A grade separation structure passes over a two-lane highway. The normal cross section of the two-lane highway has cross slopes of −2 percent from the centerline, and the roadway is on a tangent horizontal alignment. The centerline of the grade separation structure crosses the centerline of the two-lane highway at station 10 + 50 (100 m stations). At this point, the highway is in a 300 m vertical curve connecting a + 3.00 percent grade to a + 0.50 percent grade; the PVC is station 8 + 60, and the PVC elevation is 152.00 m. The grade separation structure slopes down at 1.00 percent from the left to the right of the highway as viewed by an observer facing in the direction of increasing stations and is 14 m wide. The minimum elevation of the bottom of the grade separation required to provide 5.0 m clearance over all points of the traveled way of the two-lane road is approximately:
a. 161.3 m
b. 161.5 m
c. 161.7 m
d. 161.9 m

4.39 The table below gives earthwork cross section areas for three stations.

Station	Cut End Area, ft²	Fill End Area, ft²
50 + 00	120.25	0
50 + 40	60.75	0
51 + 20	15.87	25.5

If the average end area method is used to compute the earthwork volume, the total volume of excavated material is approximately:
a. 197 yd³
b. 228 yd³
c. 248 yd³
d. 685 yd³

4.40 An older two-lane highway with a design volume of 1700 veh/day is constructed with 11-foot lanes. A bridge on this route is to be reconstructed. If the length of the bridge is 75 ft, the required minimum clear roadway width for the bridge is:
a. 22 ft
b. 26 ft
c. 30 ft
d. 32 ft

CHAPTER 5

Afternoon Exam—Transportation Engineering

5.1 Speeds of five successive vehicles on a lightly traveled rural road were measured by a radar gun as they passed an observer. The speeds were 57.6 mph, 62.4 mph, 62.1 mph, 74.4 mph, and 49.6 mph. The space-mean and time-mean speeds of these five vehicles are most nearly:
  a. space-mean speed = 61.2 mph, time-mean speed = 61.2 mph
  b. space-mean speed = 60.7 mph, time-mean speed = 61.2 mph
  c. space-mean speed = 60.2 mph, time-mean speed = 60.7 mph
  d. space-mean speed = 60.2 mph, time-mean speed = 61.2 mph

5.2 The table below gives traffic volumes for 12 hours of an average day at an intersection that is being considered for signalization.

Time Period	Major Street Volume (both approaches)	Minor Street Volume (highest volume approach)
0600–0700	750	120
0700–0800	950	180
0800–0900	800	200
0900–1000	500	140
1000–1100	450	120
1100–1200	560	80
1200–1300	620	125
1300–1400	530	120
1400–1500	580	160
1500–1600	790	190
1600–1700	990	160
1700–1800	650	100

The major street approach has two lanes, and the minor street approach has one lane. The intersection is located in a major city, and the approach speeds are about 25 mph. Which signal warrants (if any) are met?
  a. warrant 1 only
  b. warrant 2 only
  c. warrant 3 only
  d. none of warrants 1, 2, and 3

5.3 A conventional two-lane rural highway has a design speed of 100 km/h. This roadway intersects a minor road. This intersection is controlled by stop signs on the minor road approaches. The minor road is expected to be used almost exclusively by passenger cars. Grades for both roads are about 2 percent. The design intersection sight distance along the major road that is required for this intersection is:
a. 185 m
b. 240 m
c. 210 m
d. 250 m

5.4 An urban freeway has three lanes in each direction. Lane width is 12 ft, and there are 10 ft shoulders on the right side of the traveled way. Within a 6-mile segment surrounding the point being analyzed (3 miles upstream and 3 miles downstream), there are a total of seven interchanges. The estimated free-flow speed is most nearly:
a. 70 mph
b. 68 mph
c. 66 mph
d. 64 mph

5.5 Adjusted peak flow rates for the ramp junction shown in Exhibit 5.5 are 7200 pc/h for the freeway upstream of the ramp and 700 pc/h for the ramp.

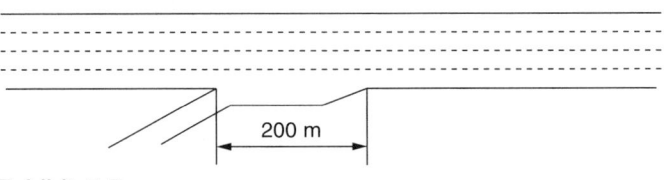

Exhibit 5.5

The free-flow speed on the ramp is 60 km/h. The estimate of the total flow entering the ramp influence area is most nearly:
a. 7200 pc/h
b. 2350 pc/h
c. 700 pc/h
d. 3050 pc/h

5.6 A gravity model of the form $T_{ij} = P_i(A_j F_{ij}/\Sigma A_j F_{ij})$ is to be used to distribute the productions from zone 1 in the table below to the other zones.

Zone	Productions	Attractions	Impedance for Trips From Zone
1	1000	600	N/A
2	500	400	12
3	700	1200	30
4	900	700	16
5	300	500	22

The number of trips from zone 1 to zone 3 is most nearly:
a. 230
b. 570
c. 440
d. 685

5.7 In the mass diagram shown in Exhibit 5.7a, the limit of economic haul is 2000 ft. Points for which the LEH spans the various loops are shown in the diagram. For the job shown, the total amount of borrow is approximately:
a. 10,000 yd^3
b. 15,000 yd^3
c. 25,000 yd^3
d. 30,000 yd^3

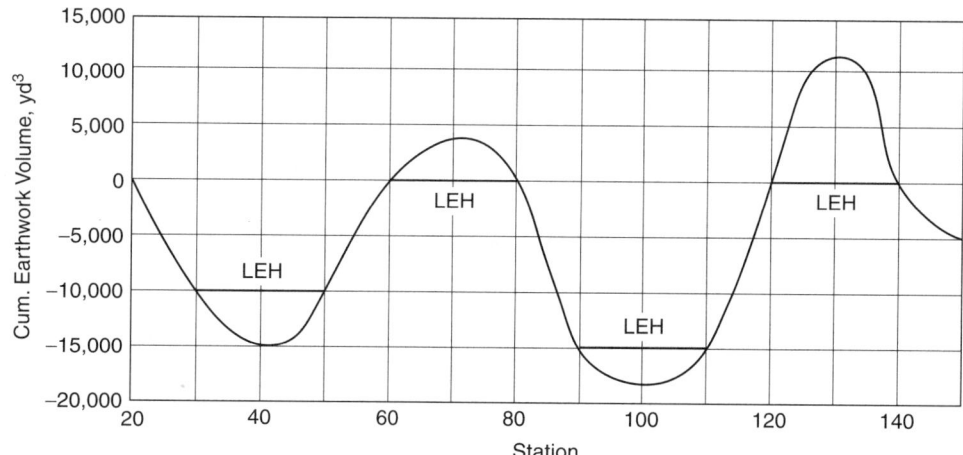

Exhibit 5.7a

5.8 An asphalt concrete pavement is being designed for a roadway with a design ESAL value of 3×10^6. The pavement structure is to consist of asphalt concrete with an elastic modulus of 400,000 psi, a granular base with an elastic modulus of 24,000 psi, and a granular subbase with an elastic modulus of 12,000 psi. The elastic modulus of the subgrade is 8000 psi. The reliability level to be used in the design is 90 percent. The difference between the initial serviceability index and terminal serviceability index is assumed to be 1.7. The combined standard error of the traffic prediction and performance prediction is assumed to be 0.45, which is typical of flexible pavements. The structural number of the asphalt surface is most nearly:
a. 2.7
b. 4.5
c. 3.3
d. 3.8

5.9 A four-lane undivided highway in a rural area (design speed = 100 km/h) has a 1000 m radius horizontal curve with a superelevation rate of 5.3 percent. Lane width is 3.6 m. The minimum length of superelevation runoff is most nearly:
a. 80 m
b. 55 m
c. 65 m
d. 60 m

5.10 A vertical curve connecting a −1.5 percent grade with a +2.3 percent grade has a PI at station 20 + 44 and elevation 325.00 ft above sea level. The roadway must clear the top of a pipe located at station 18 + 94 and elevation 328.50 ft above sea level by 2.2 ft. The shortest vertical curve that can be used is approximately:
a. 1440 ft
b. 2500 ft
c. 870 ft
d. 1250 ft

5.11 A vehicle involved in an accident is estimated to have been traveling at 20 mph at the time of impact. Skid marks are measured and found to be 150 feet long. The roadway is on a +1.5 percent grade, and the coefficient of friction is estimated to have been 0.30. The speed of the vehicle at the time the skid began was approximately:
a. 55 mph
b. 65 mph
c. 70 mph
d. 45 mph

5.12 Characteristics of a roadway section are given in the table below.

Section	Length, mi	Injury Accidents	AADT
1	1.5	12	10,000
2	3.0	35	14,000
3	1.8	15	17,000

The injury accident rate for the whole section is most nearly:
a. 1.75 per million vehicle-miles
b. 1.95 per million vehicle-miles
c. 2.05 per million vehicle-miles
d. 2.00 per million vehicle-miles

5.13 The shoulders of a two-lane rural highway are to be widened by 2 ft on each side. This improvement is expected to reduce the number of related accidents by 32 percent. Over the past 3 years, the number of related accidents has been 15, 18, and 12 respectively. The current ADT is 6000. The expected ADT in the after period used for comparison is 7500. The reduction in the annual number of accidents that is expected to result from implementation of this countermeasure is most nearly:
a. 4
b. 5
c. 6
d. 7

5.14 Data are given below for each lane group of an intersection.

Lane group	NBLT	SBLT	NB	SB	EB	WB
Saturation flow	1750	1765	3500	3450	1800	1820
Volume	298	265	1190	1070	486	564
v/s	0.17	0.15	0.34	0.31	0.27	0.31

Lost time is equal to 3 seconds per change interval. The optimum cycle as determined by Webster's method is most nearly:
a. 60 s
b. 80 s
c. 90 s
d. 100 s

5.15 An urban bus route has the following estimated travel times (including stops):

Northbound
Main Street (southern terminus) to Acme Drive — 35 min
Acme Drive to Madison Avenue — 25 min
Madison Avenue to Jones Street (northern terminus) — 40 min

Southbound
Jones Street to Madison Avenue — 42 min
Madison Avenue to Acme Drive — 26 min
Acme Drive to Main Street — 33 min

Layovers of at least 10 min each are required at both ends of the route. Headways are 15 min. The minimum number of vehicles required to operate the route is:
a. 10
b. 15
c. 20
d. 25

5.16 A vertical curve is to be designed to connect a +2.0 percent grade to a −4.0 percent grade. The minimum length of vertical curve required to provide stopping sight distance for a design speed of 60 mph is approximately:
a. 900 ft
b. 950 ft
c. 1000 ft
d. 1100 ft

5.17 The following table summarizes the results of a spot speed study.

In the absence of a statutory speed limit, the appropriate speed limit for this roadway is:
a. 45 mph
b. 50 mph
c. 55 mph
d. 60 mph

Speed Class			Number of Observations	Cumulative Frequency	Cumulative Percent
Lower Bound (mph)	Midpoint (mph)	Upper Bound (mph)			
27.6	30.0	32.5	3	3	2.4
32.6	35.0	37.5	7	10	8.0
37.6	40.0	42.5	15	25	20.0
42.6	45.0	47.5	22	47	37.6
47.6	50.0	52.5	35	82	65.6
52.6	55.0	57.5	27	109	87.2
57.6	60.0	62.5	14	123	98.4
62.6	65.0	67.5	2	125	100.0
Total			125		

5.18 A 600 m horizontal curve (with no spirals) connects a tangent with a bearing of N 75° E to a tangent with a bearing of N 30° E. The station of the PI of the tangents is 1 + 127. The station of the PC is most nearly:
a. 0 + 249
b. 0 + 878
c. 1 + 127
d. 1 + 376

5.19 During the peak 4 hours of the day, demand for a proposed parking lot is expected to include 375 shoppers, who will park for an average of 30 minutes each, and 50 employees, who will remain parked for the entire 4 hours. If parking efficiency is expected to be 0.85, the required number of spaces is approximately:
a. 70
b. 85
c. 100
d. 115

5.20 Volume counts are taken on a roadway, and the average daily traffic (ADT) for the month of March is found to be 2500 veh. The monthly expansion factor for March is 0.95. The annual average daily traffic (AADT) is approximately:
a. 2125 veh
b. 2375 veh
c. 2630 veh
d. 2940 veh

5.21 A proposed discount store is planned with 120,000 ft^2 of gross floor area. An estimate of the number of trips generated during the peak hour of adjacent street traffic on a weekday is needed as part of a traffic impact study. The local planning commission has established a standard that trip generation estimates shall provide that the probability that the estimate will be exceeded will be no greater than 0.15. ITE (1998) states that for the peak hour of adjacent street traffic on a weekday, the average trip generation rate for such stores is 4.24 per 1000 ft^2 of gross floor area with a standard deviation of 2.23 per 1000 ft^2 of gross floor area. The required estimate of

the number of trips generated during the peak hour of adjacent street traffic on a weekday is approximately:
a. 200
b. 510
c. 775
d. 1050

5.22 A traffic signal has a 90 s cycle. Effective green time for pedestrians crossing the major street is 18 s. The level of service for pedestrians crossing this street is:
a. B
b. C
c. D
d. E

5.23 An arterial street has cross streets spaced as follows: A Street to B Street, 90 m; B Street to C Street, 140 m; and C Street to D Street, 120 m. A one-way signal coordination scheme is proposed, using a 60 s cycle and a speed of progression of 50 km/h. The offset between A Street and D Street should be approximately:
a. 7 s
b. 10 s
c. 25 s
d. 60 s

5.24 The following table gives two-way traffic volume and signal spacing for four arterial streets.

Street	Two-way Volume, veh/h	Signal Spacing, ft
Elm Street	700	600
Ash Street	450	1200
Poplar Street	400	900
Walnut Street	800	450

Coordinated signals should be considered for:
a. Elm Street and Ash Street
b. Elm Street and Walnut Street
c. Poplar Street and Walnut Street
d. Poplar Street and Ash Street

5.25 A driver is traveling at 30 mph on a +1.5 percent grade when a child suddenly runs into the roadway. Assuming a reaction time of 2.5 s and a deceleration rate of 11.2 ft/s^2, the distance the vehicle will travel before stopping is most nearly:
a. 190 ft
b. 230 ft
c. 270 ft
d. 300 ft

5.26 A vehicle traveling at 100 km/h traverses a horizontal curve with a radius of 500 m. If the superelevation rate is 0.09, the side friction is most nearly:
a. 0.05
b. 0.06
c. 0.07
d. 0.08

5.27 A soil has a void ratio (e) of 0.8, moisture content (w) of 15 percent, and specific gravity of soil solids (G_s) of 2.68. The degree of saturation (S) is nearly equal to:
a. 30%
b. 40%
c. 50%
d. 60%

5.28 A moist soil sample obtained from the field has the following characteristics:

Mass of moist soil = 432 g
Mass of dry soil = 320 g
Liquid limit, LL = 48%
Plastic limit, PL = 21%

The liquidity index of the soil is nearly equal to:
a. 0.5
b. 0.6
c. 0.7
d. 0.8

5.29 The void ratio (e) and the coefficient of permeability (k) of a clay soil are as follows:

e	k, cm/s
1.2	0.6×10^{-7}
1.52	1.519×10^{-7}

Given: $k \propto \dfrac{e^n}{1+e}$

The coefficient of permeability at a void ratio of 1.4 will be nearly equal to:
a. 0.7×10^{-7} cm/s
b. 0.9×10^{-7} cm/s
c. 1.1×10^{-7} cm/s
d. 1.3×10^{-7} cm/s

5.30 Water flows through a permeable layer as shown in Exhibit 5.30. The coefficient of permeability (k) of the permeable layer is 0.03 in./s. The rate of flow of water through the permeable layer is nearly equal to:
a. 1×10^{-3} ft^3/s/ft
b. 1.5×10^{-3} ft^3/s/ft
c. 1.95×10^{-3} ft^3/s/ft
d. 2.5×10^{-3} ft^3/s/ft

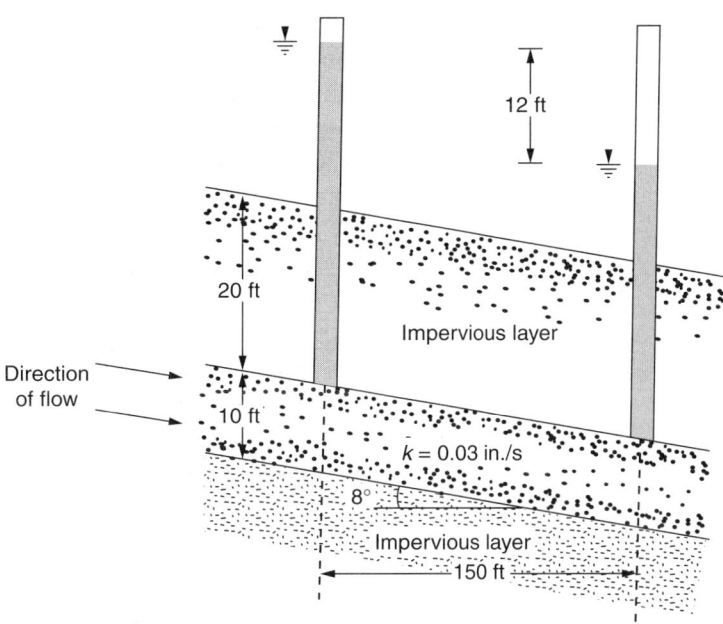

Exhibit 5.30

5.31 During a field exploration, rock coring was done for a total length of 2.5 ft. The rock core recovered was as follows:

Two 6 in. length pieces
Two 8 in. length pieces

The recovery ratio is nearly equal to:
a. 60%
b. 70%
c. 82%
d. 93%

5.32 A soil has the following characteristics: percent passing No. 40 U.S. sieve = 71, percent passing No. 200 U.S. sieve = 52, liquid limit = 31, plastic limit = 12. The classification of the soil according to the AASHTO system, including group index, is:
a. A-6(6)
b. A-7-6(15)
c. A-4(8)
d. A-5(8)

5.33 Given an overflow rate of 20 m/day and a flow of 0.2 m^3/s, the detention time in a 2.5 m deep sedimentation basin with a length:width ratio of 2:1 is most nearly:
a. 40 minutes
b. 100 minutes
c. 180 minutes
d. 220 minutes

5.34 Consider steady and uniform flow in a prismatic channel. The channel is trapezoidal in cross section, with a base of width 5 ft and sides sloped at an angle of 30° from the horizontal. The channel slopes 10 feet over its 0.5-mile length. Where is the velocity the greatest in the channel?
   a. at the upstream end
   b. at the downstream end
   c. 2/3 of the distance from the upstream end
   d. None of the above

5.35 Which of the following has the most efficient hydraulic section?
   a. semicircle
   b. triangle with 45° included angle
   c. trapezoid with 45° angles
   d. trapezoid with 30° angles, as measured from the horizontal

5.36 The flow immediately upstream of a hydraulic jump is:
   a. always critical
   b. always subcritical
   c. always supercritical
   d. Depending on specific channel conditions, may be any one of the above.

5.37 Catchments X and Y drain to stormwater Inlets X and Y, respectively (Exhibit 5.37a). The 18-inch concrete pipe connecting Inlets X and Y is 1000 feet long and is laid at a slope of 2 percent. For the Intensity-Duration curve for a 10-year return period (Exhibit 5.37b), the design intensity to be used in sizing the pipe draining the combined flow of both catchments is most nearly:
   a. 1 in./hr
   b. 2 in./hr
   c. 4 in./hr
   d. 6 in./hr

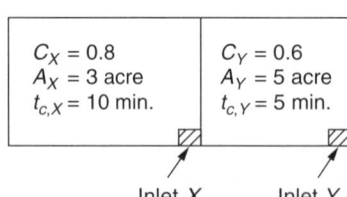

Exhibit 5.37a

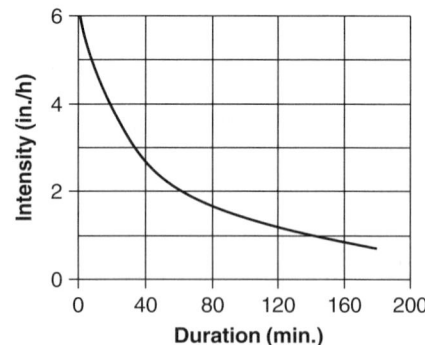

Exhibit 5.37b

5.38 The flow in a 12-inch concrete pipe when it flows full is 1 cfs. The flow and velocity in that same pipe, when flowing half-full, is most nearly:
   a. 1 cfs and 1.3 fps
   b. 1 cfs and 0.65 fps
   c. 0.5 cfs and 1.3 fps
   d. 0.5 cfs and 0.65 fps

5.39 A water main is to be designed to serve a proposed 100-home residential subdivision on 50 acres. Which of the following statements is true?
   a. The required fire flow is nearly the same as the average day demand.
   b. The required fire flow is between 2 times and 5 times the average day demand.
   c. The required fire flow is between 5 times and 100 times the average day demand.
   d. The required fire flow is more than 100 times the average day demand.

5.40 A 50-acre catchment containing cropland is converted to a "big-box" retail center. Which of the following statements is true concerning the impact on the runoff hydrograph?
   a. The postdevelopment hydrograph peak will occur sooner than the predevelopment hydrograph peak.
   b. The postdevelopment hydrograph peak will be higher than the predevelopment hydrograph peak.
   c. The area under the postdevelopment hydrograph will be greater than the area under the predevelopment hydrograph.
   d. All of the above

Afternoon Exam—Construction Engineering

6.1 ABC Construction Company is the low bidder at $500,000 on a building project. The company has provided the typical bid bond. After award of the contract, the company decides to withdraw its bid. The second-lowest bid was $520,000. How much is the surety company required to reimburse the client for the failure of ABC to sign?
   a. $500,000
   b. $50,000
   c. $20,000
   d. $5,000

6.2 A contractor has bid a lump sum of $50,000 (direct costs) to install 200 linear feet of pipe, plus $5,000 overhead and profit (O&P). The client wants the contractor to quote a unit price for the work. What should the contractor charge per linear foot?
   a. $200/ft
   b. $250/ft
   c. $275/ft
   d. $300/ft

6.3 During the summer break on a school renovation project, a crew of a mason and a laborer lay 100 modular-size concrete bricks, $4'' \times 4'' \times 8''$, per hour under normal conditions. Before the school year begins, a total of 56,000 bricks have been installed. After this point, the owner limits the access to the construction area, and the contractor's crew must lay 32,000 bricks on the following 50 working days. Using the Measured Mile method (MM), determine the contractor's compensable delay due to the owner's interference. (Assume that the mason's and the laborer's salaries are $12/hr and $8/hr, respectively. The payroll taxes and fringe benefits are calculated at 40 percent of the labor direct cost.)
   a. $1120
   b. $1600
   c. $2240
   d. $3200

6.4 Determine the volume of soil (yd³) needed for fill using the average end-area method. The following information is given:

- Cross-sectional areas at both ends are embankment and are trapezoids.
- Height of fill at one end is 5 feet.
- Height of fill at the other end is 7 feet.
- Cross section has a 32 ft horizontal roadway with 2:1 sideslopes.
- Distance between cross sections is 80 feet.

a. 784 cubic yards
b. 788 cubic yards
c. 1576 cubic yards
d. 21,280 cubic yards

6.5 A 40 × 40 foot raised parking pad requires fill. The heights of fill at the four corners are 2.5′, 3.2′, 2.8′, and 3.5′. How much fill is required (do not calculate the amount of material for side slopes)?
a. 1600 cubic yards
b. 533.3 cubic yards
c. 177.8 cubic yards
d. 400 cubic yards

6.6 Determine the volume of soil (yd³) using the prismoidal method. The following information is given:

- Cross-sectional areas at both ends are embankment and are trapezoids.
- Height of fill at one end is 5 feet.
- Height of fill at the other end is 7 feet.
- Cross section has a 32 ft horizontal roadway with 2:1 sideslopes.
- Distance between cross sections is 80 feet.

a. 784 cubic yards
b. 788 cubic yards
c. 3529 cubic yards
d. 21,173 cubic yards

6.7 A triangular spoil bank is 20′ 5″ wide, 10′ 0″ high, and 45′ 9″ long. A conical spoil pile has a base diameter of 30′ 11″ and a height of 15′ 0″. Determine the combined volume (yd³).
a. 312 yd³
b. 382 yd³
c. 485 yd³
d. 8424 yd³

6.8 5,000 cubic yards of compacted soil will be used to fill a depression. The shrinkage factor from bank measure to the final compacted state is 0.75. The swell factor from bank measure to loose soil is 30 percent for transporting the soil. If a truck can hold 12 cubic yards of loose measure soil, determine the number of truckloads needed.
a. 313 truckloads
b. 417 truckloads
c. 556 truckloads
d. 722 truckloads

6.9 Determine the area (ft²) for the cross section of highway shown in Exhibit 6.9a.

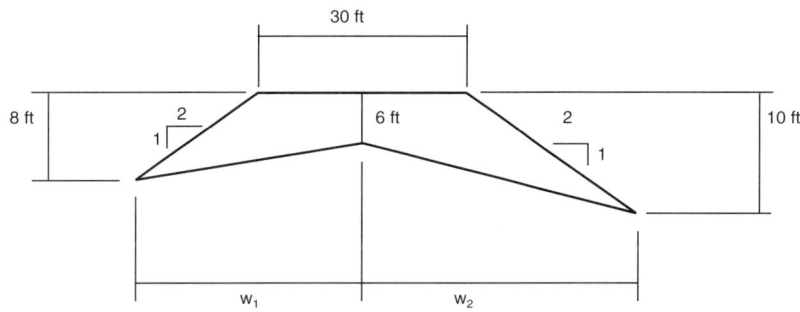

Exhibit 6.9a

a. 333 ft²
b. 389 ft²
c. 468 ft²
d. 666 ft²

6.10 A triangular spoil bank is 10′ 6″ wide, 5′ 0″ high, and 40′ 0″ long. The swell factor from spoil bank to loose soil is 15 percent for transporting the soil. If a truck can hold 12 cubic yards of loose measure soil, determine the number of truckloads needed to move the pile of soil.
a. 3 truckloads
b. 4 truckloads
c. 58 truckloads
d. 101 truckloads

6.11 A concrete foundation for a rectangular building consists of a footing that is 3 ft wide by 1 ft deep and a 1 ft wide by 5 ft high wall. The rectangular building has outside dimensions of 40 × 30 ft. The outside of the wall has the same outside dimensions as the building, and the wall is centered on the footing. The volume of concrete (yd³) needed for the foundation, adding in 5 percent for waste, is most nearly
a. 40 yd³
b. 41 yd³
c. 43 yd³
d. 44 yd³

6.12 A reinforced concrete slab is 27′ 8″ by 19′ 8″. #4 rebars are to be placed 8″ center-to-center each way. The #4 rebars come in 20 ft lengths. The rebar must have a 2″ cover at the ends of the slab. An overlap splice of 2 ft is required along the 27′ 8″ length. Determine the total number of rebars needed.
a. 87 rebars
b. 102 rebars
c. 45 rebars
d. 68 rebars

6.13 Your lab technician is performing a standard Proctor test and wants you to check the results. The technician has used a standard mold that weighs 9.2 lbs empty. After compaction, the mold and soil weigh 12.96 lbs. A sample of this soil weighs 112 g and 102 g after drying. What is the dry density of this sample?
   a. 102.7 pcf
   b. 112.8 pcf
   c. 109.8 pcf
   d. 9.8%

6.14 A road project requires a dry density of 110 pcf at a moisture content of 10 percent. Tests indicate that the current moisture content is at 8 percent. How much water should be added to the 6" lift per station if the road width is 32 ft?
   a. 4.2 gal/yd^2
   b. 1.2 gal/yd^2
   c. 120 gal/yd^2
   d. 2.0 gal/yd^2

6.15 Your company will be excavating a trench with a 4 ft wide bottom, 10 ft deep, in a cohesive soil with an unconfined compressive strength of 1 tsf. How wide should the top of the trench cut be?
   a. 20'
   b. 12'
   c. 24'
   d. No slope required

6.16 A wheeled loader with a 2.5 yd^3 bucket moves loose material up an effective grade of 15 percent for 200 ft. The loader works 45 minutes per hour. Exhibit 6.16a summarizes loader cycle times and Exhibit 6.16b shows typical loader travel times. The production for the loader described is most nearly:

Exhibit 6.16a *Loader Cycle Times*

Loading Conditions	Basic Cycle Time (minutes)	
	Wheeled Loader	Tracked Loader
Loose Material	0.35	0.30
Average Material	0.50	0.35
Hard Material	0.65	0.45

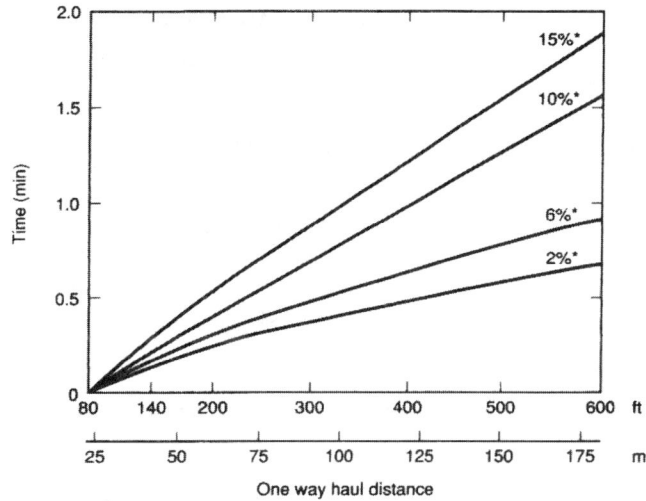

*Effective grade

Exhibit 6.16b *Typical Loader Travel Times*

a. 120 lcy/hr
b. 147 lcy/hr
c. 132 lcy/hr
d. 176.5 lcy/hr

6.17 A motor grader is performing fine grading of a road subbase. It operates at 3.5 mph and requires two passes to complete the grading. The effective blade width is 8 ft, and efficiency is 0.60. Determine what length of a 40′ wide roadbed it can fine grade each hour.
a. 9.5 sta/hr
b. 13.3 sta/hr
c. 10.8 sta/hr
d. 19.0 sta/hr

6.18 A 400 hp scraper operates 1000 hours per year. The cost of the scraper is $170,000, and the tires cost $10,000 per set. Tire life is 1000 hours. Fuel costs $2.60 per gallon. The operator makes $20 per hour with 25 percent fringes. The scraper works under average conditions. Exhibits 6.18a–c provide general estimating data. A common equation for estimating operating costs for tires is

Tire operating cost = 1.15 × (cost of set of tires / expected tire life [hrs])

Exhibit 6.18a Fuel Consumption (gal/hr/horsepower)

Equipment	Load Conditions		
	Low	Average	Severe
Compactor	0.040	0.050	0.060
Excavator	0.035	0.040	0.050
Wheeled Loader	0.025	0.035	0.045
Scraper	0.025	0.035	0.045
Dozer	0.030	0.040	0.050
Grader	0.025	0.035	0.045
Off-Road Truck	0.015	0.020	0.030

Exhibit 6.18b Service Costs (Percent of Hourly Fuel Cost)

Operating Conditions	Service Cost Factor
Favorable	20% of fuel cost
Average	33% of fuel cost
Severe	50% of fuel cost

Exhibit 6.18c Lifetime Repair Costs as a Percent of Initial Cost

Equipment	Favorable	Average	Severe
Shovels/Hoes	50%	70%	90%
Loaders/Dozers	60%	75%	90%
Scrapers	80%	90%	100%
Dump Trucks	70%	80%	90%

The first-year operating cost for the scraper is
a. $70.07/hr
b. $83.57/hr
c. $95.57/hr
d. $106.34/hr

6.19 On a residential development project, the site subcontractor has accumulated 1,480 cubic meters (m^3) of excavated clay material at the job site. Using a wheel loader and 5 trucks with an 18 cubic meter capacity, determine the amount of time needed to dispose all the soil material.

The following information is given:

- Loading time: 2.0 minutes
- Amount of time from the loading point to the unloading point outside the project site: 15 minutes
- Unloading time: 1.5 minutes
- Amount of time from the unloading point back to the project site: 11.5 minutes
- Efficiency adjustment = 45 minutes
- Clay pounds per cubic yards-bank = 2,960
- Clay pounds per cubic yards-loose = 2,130

a. 8.00 hours
b. 9.50 hours
c. 14.25 hours
d. 15.50 hours

6.20 Given the CPM diagram in Exhibit 6.20, the activities on the critical path are
a. A-B-E-G
b. A-D-G
c. A-C-F-G
d. A-B-F-G

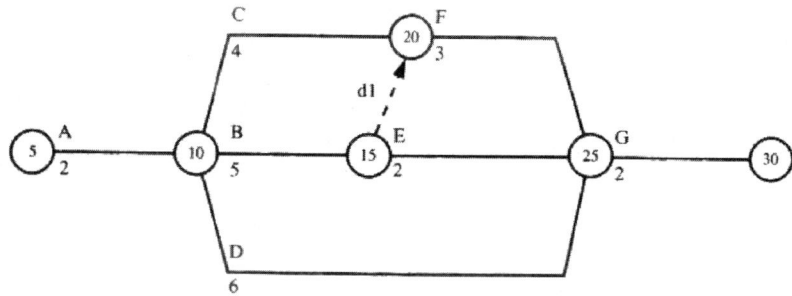

Exhibit 6.20

6.21 Using the information contained in Exhibit 6.21, determine the total project duration.

Exhibit 6.21

Activity	Duration (Days)	Immediate Predecessor
A	8	–
B	11	–
C	3	A
D	6	B, C
E	2	B
F	1	B
G	3	E
H	5	D, G
I	2	E, F
J	2	I

a. 19 days
b. 21 days
c. 22 days
d. 23 days

6.22 Using the diagram shown in Exhibit 6.22a, determine the project expected duration (t_e).

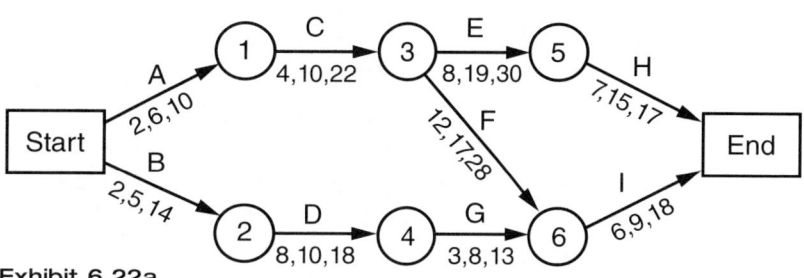

Exhibit 6.22a

a. 35
b. 40
c. 45
d. 50

6.23 Using the information Exhibit 6.23a, determine the Early Finish (EF) time for activity G.

Exhibit 6.23a

Activity	Duration (Days)	Immediate Predecessor
A	5	–
B	3	–
C	6	–
D	10	A
E	12	A,B,C
F	11	C
G	5	D,E
H	7	E
I	9	F
J	10	G,H

a. 18 days
b. 23 days
c. 25 days
d. 26 days

6.24 A contractor hires you as a consultant on a residential development with a signed contract with the owner for the amount of $7,500,000. Your client's CPM has determined that the owner is responsible for a 60-day period due to additional work requested beyond the original scope of work. The total duration of the project, including the performance of the additional work, was 450 days. During that same period, the contractor had four projects underway with a total company billing of $25,000,000. The total home office overhead for that period was $1,500,000. Determine the amount of extended office overhead that should be allocated on a claim against the owner for the delay of the project.
a. $50,000
b. $60,000
c. $450,000
d. $600,000

6.25 During the execution of an office renovation project, your boss wants you to predict the estimate at completion (EAC) for the windows installation trade. Your budget for this trade is $250,000, but to date, your actual cost is $200,000. Your schedule shows that 100 percent of this activity should be finished by today, but your latest status report indicates that only 60 percent has been completed. What is the estimate at completion (EAC), assuming that the delays encountered will continue on the remaining part of the installation?
a. $192,000
b. $320,000
c. $336,000
d. $368,000

6.26 The latest status report shows the following information about a $250,000 project:

- Sixty-five percent of the project has been completed.
- The actual cost of the work performed is $130,000.
- Eighty-five percent of the project should be completed as of today.

What is the Schedule Variance (SV) for the project?
a. $50,000
b. −$32,500
c. −$50,000
d. −$72,500

6.27 Given the network diagram in Exhibit 6.27, determine the maximum amount of time that activity B can be delayed without affecting the total duration of the project.

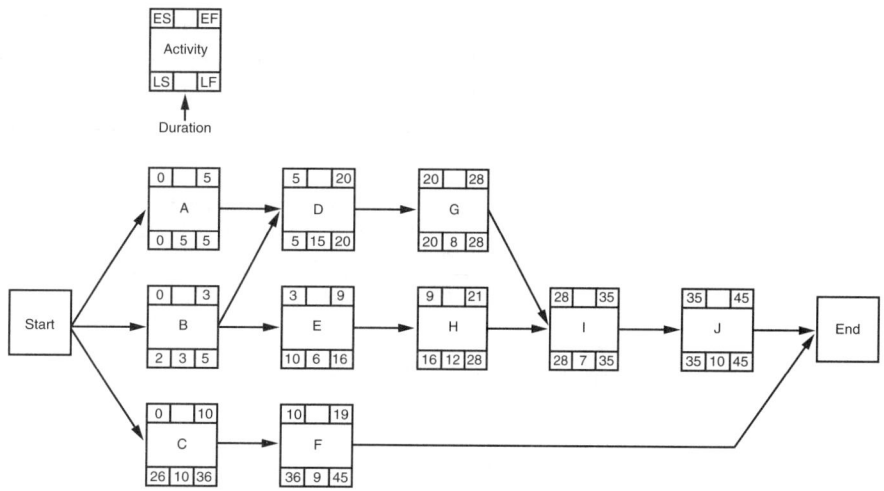

Exhibit 6.27

a. 2
b. 3
c. 5
d. 10

6.28 Using the bar chart in Exhibit 6.28a, perform resource leveling and determine the minimum number of constant resources (painters) required to complete the project in eight days. Assume each task can be performed independently of the other tasks.

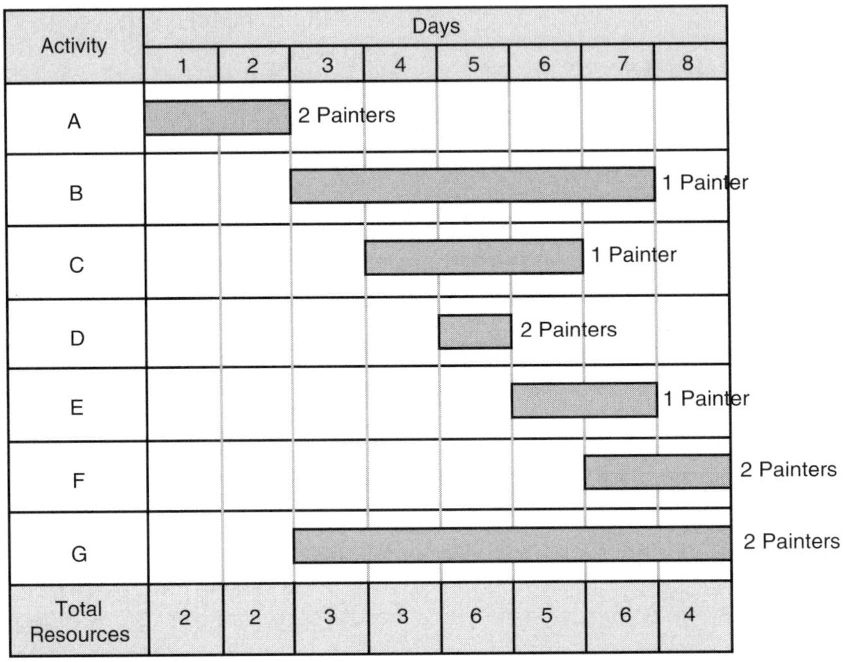

Activity	Days							
	1	2	3	4	5	6	7	8
A	\multicolumn{2}{l}{2 Painters}						
B			\multicolumn{6}{l}{1 Painter}
C				\multicolumn{4}{l}{1 Painter}	
D					\multicolumn{2}{l}{2 Painters}		
E						\multicolumn{2}{l}{1 Painter}	
F							\multicolumn{2}{l}{2 Painters}
G			\multicolumn{6}{l}{2 Painters}
Total Resources	2	2	3	3	6	5	6	4

Exhibit 6.28a

a. 2
b. 3
c. 4
d. 5

6.29 The following relationships exist between activities for a project. No preceding activities for A and D. Activities A and D precede C and E. Activity E precedes B. Activity C precedes F. Activities B and F precede G.

The project's Gantt chart in Exhibit 6.29 was updated at the end of ten days showing what was completed through the tenth day. Determine the new critical path of the project. The numbers on top of the Gantt chart refer to working days.

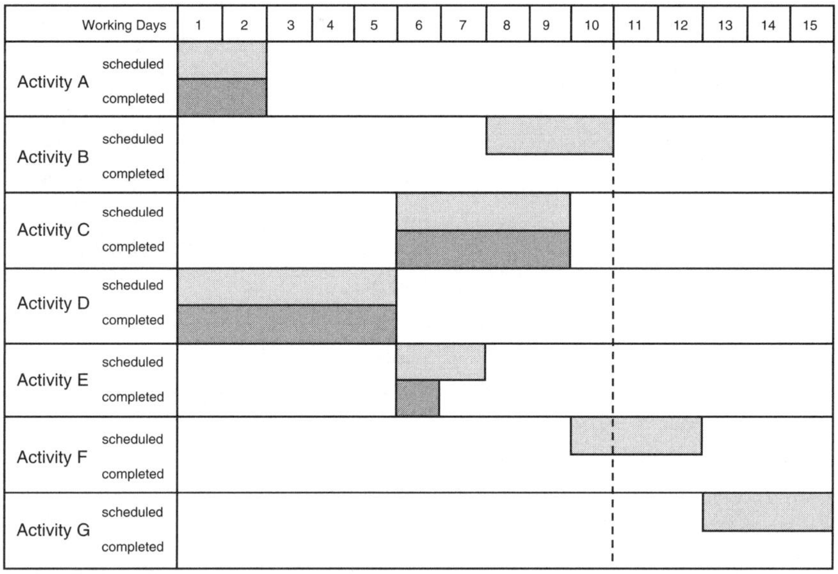

Exhibit 6.29

a. A-B-C-D
b. A-D-E-C
c. D-E-B-G
d. D-C-F-G

6.30 The following relationships exist between activities for a project. No preceding activities for A and C. Activity A precedes B. Activities B and C precede D. Activity D precedes E.

The project's Gantt chart in Exhibit 6.30 was updated at the end of seven days showing what was completed through the seventh day. Determine the duration of the project. The numbers on top of the Gantt chart refer to working days.

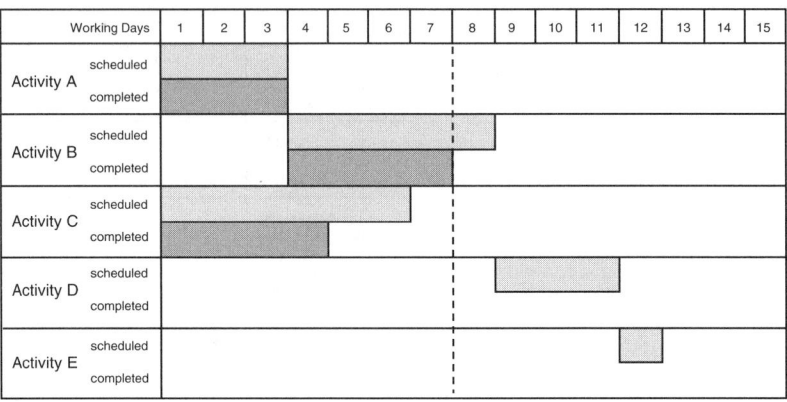

Exhibit 6.30

a. 10 days
b. 12 days
c. 13 days
d. 14 days

6.31 Determine the expected duration of the project using PERT for the precedence diagram shown in Exhibit 6.31a. All durations are shown in days.

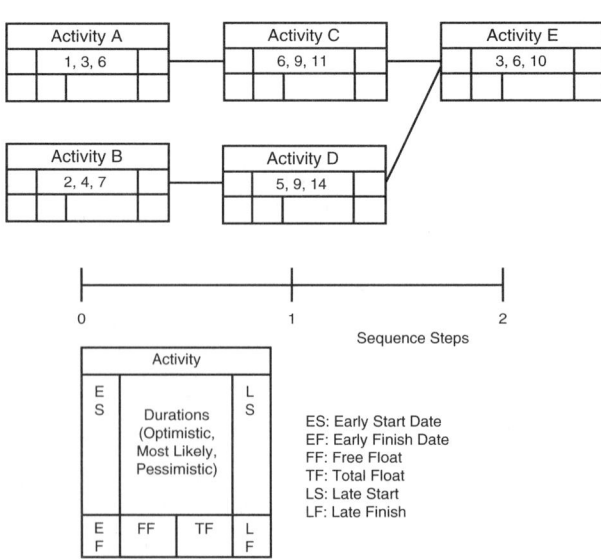

Exhibit 6.31a

a. 18.2 days
b. 19.2 days
c. 20.6 days
d. 19.6 days

6.32 Using PERT, determine the variance of the critical path for the project shown in Exhibit 6.32a. All durations are shown in days.

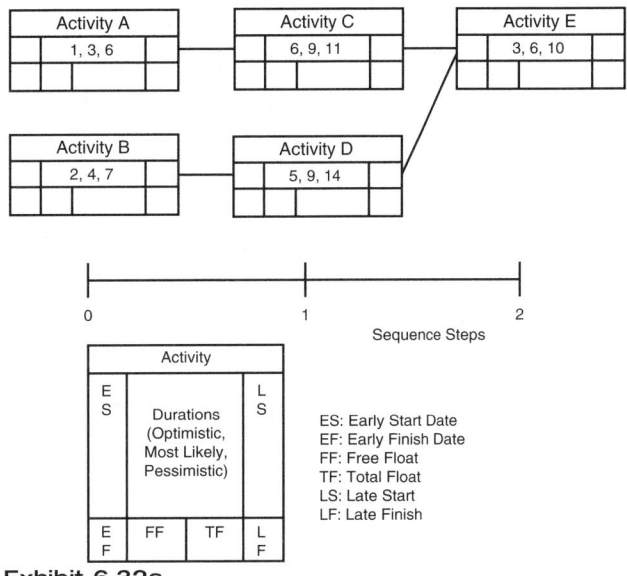

Exhibit 6.32a

a. 4.3 days
b. 3.5 days
c. 21 days
d. 18.5 days

6.33 Determine the total float in activity E of Exhibit 6.33a. All durations are shown in days.

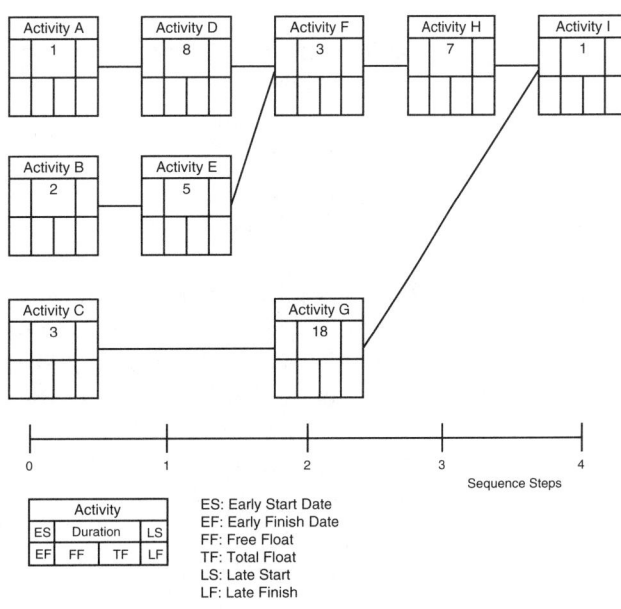

Exhibit 6.33a

a. 2 days
b. 4 days
c. 0 days
d. 1 day

6.34 Determine the critical path for the schedule shown in Exhibit 6.34a. All durations are shown in days.

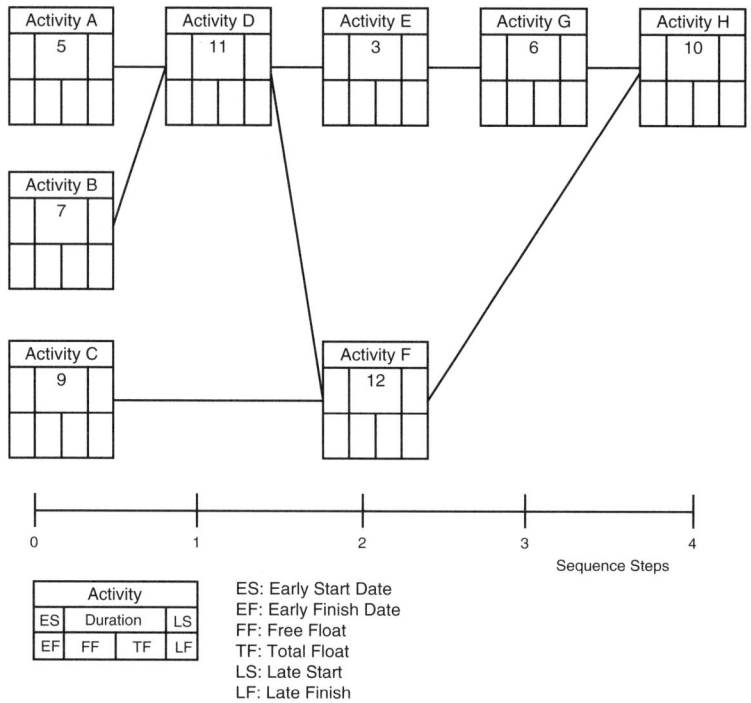

Exhibit 6.34a

a. B-D-F-G-H
b. A-D-E-G-H
c. C-F-G-H
d. B-D-E-G-H

6.35 For the schedule shown in question 6.34, determine the duration of the project.
a. 47 days
b. 46 days
c. 35 days
d. 37 days

6.36 A project was completed at the end of December 2003 for a cost of $700,000. Since then, the following cost-of-living increases have occurred: 5.5% in 2004, 4% in 2005, 5% in 2006, 5% in 2007, and 6% in 2008. To build that same project for completion at the end of May 2008 would cost most nearly:
a. $898,000
b. $868,000
c. $867,000
d. $897,000

6.37 A piece of equipment costs $150,000 and has a life expectancy of five years. At the end of five years, the salvage value is $20,000. Using double-declining-balance depreciation, determine the book value of the equipment after the third year.
  a. $32,400
  b. $28,080
  c. $72,000
  d. $21,600

6.38 A piece of equipment costs $38,000 and has a life expectancy of four years. At the end of four years, the salvage value is $7,000. Using sum-of-years-digits depreciation, determine the book value of the equipment after the third year.
  a. $23,250
  b. $14,750
  c. $6,200
  d. $10,100

6.39 A piece of equipment is purchased for $80,000. A down payment of $5,000 is made. A loan is to be taken out for five years at an interest rate of 7 percent. Determine the monthly payments to the nearest whole dollar.
  a. $1753
  b. $1250
  c. $1338
  d. $1485

6.40 If $500 per month is deposited in a bank with 3 percent annual interest, determine the balance in the account after 15 years.
  a. $90,000
  b. $92,700
  c. $113,500
  d. $114,900

CHAPTER 7

Solutions

OUTLINE

SOLUTIONS TO CHAPTER 1 PROBLEMS   83

SOLUTIONS TO CHAPTER 2 PROBLEMS   97

SOLUTIONS TO CHAPTER 3 PROBLEMS   112

SOLUTIONS TO CHAPTER 4 PROBLEMS   126

SOLUTIONS TO CHAPTER 5 PROBLEMS   141

SOLUTIONS TO CHAPTER 6 PROBLEMS   149

SOLUTION SUMMARIES   165

SOLUTIONS TO CHAPTER 1 PROBLEMS

1.1 c. The factored load is given by ACI Eq. (9-2) as

$$w_u = 1.2 \times w_D + 1.6 \times w_L$$
$$= 1.2 \times 3 + 1.6 \times 7$$
$$= 14.8 \text{ kips/ft}$$

The factored moment is

$$M_u = w_u \ell^2/8$$
$$= 14.8 \times 10^2/8$$
$$= 185.0 \text{ kip-ft}$$

For a tension-controlled section with $\varepsilon_t = 0.005$ and $\varepsilon_c = 0.003$:

The depth of the neutral axis is

$$c_t = 0.375 d_t$$
$$= (0.375)(18)$$
$$= 6.75 \text{ in.}$$

The depth of the stress block is

$$a_t = \beta_1 c_t$$
$$= (0.85)(6.75)$$
$$= 5.74 \text{ in.}$$

The force in the stress block is

$$C_t = 0.85 f'_c b a_t$$
$$= (0.85)(3000)(12)(5.74)$$
$$= 175{,}644 \text{ pounds}$$

and

$$C_t = T = A_s f_y$$

The reinforcement area required is

$$A_s = C_t/f_y = 175{,}644/60{,}000$$
$$A_s = 2.93 \text{ in.}^2$$

Four No. 8 bars provide an area of

$$A_s = (4)(0.79)$$
$$= 3.16 \text{ in.}^2$$
$$> 2.93 \text{ in.}^2 \ldots \text{ satisfactory}$$

1.2 b.

ASD Option

The stiffness ratios are given by AISC Part 16, Fig. C-C2.4, as

$$G = \sum (I_c/L_c) \Big/ \sum (I_b/L_b)$$
$$G_B = (541/15)/(612/30)$$
$$= 1.77$$
$$G_A = 10 \ldots \text{ pinned end}$$

The effective length factor is obtained from AISC Part 16, Fig. C-C2.4, as

$$K_x = 2.1$$

The effective length about the x-axis is

$$K_x L = 2.1 \times 15$$
$$= 31.5 \text{ ft}$$

The effective length about the y-axis is

$$K_y L = 1 \times 15$$
$$= 15 \text{ ft}$$

The slenderness ratio about the x-axis is

$$K_x L/r_x = 31.5 \times 12/5.89$$
$$= 64.2$$

The slenderness ratio about the y-axis is

$$K_y L/r_y = 15 \times 12/1.92$$
$$= 93.75 \ldots \text{governs}$$

From AISC Part 4, Table 4-1, the allowable axial load (for $K_y L = 15'$) is

$$P = 246 \text{ kips}$$

LRFD Option

The stiffness ratios are given by AISC Part 16, Fig. C-C2.4, as

$$G = \sum (I_c/L_c) \Big/ \sum (I_b/L_b)$$
$$G_B = (541/15)/(612/30)$$
$$= 1.77$$
$$G_A = 10 \ldots \text{pinned end}$$

The effective length factor is obtained from AISC Part 16, Fig. C-C2.4, as

$$K_x = 2.1$$

The effective length about the x-axis is

$$K_x L = 2.1 \times 15$$
$$= 31.5 \text{ ft}$$

The effective length about the y-axis is

$$K_y L = 1 \times 15$$
$$= 15 \text{ ft}$$

The slenderness ratio about the x-axis is

$$K_x L/r_x = 31.5 \times 12/5.89$$
$$= 64.2$$

The slenderness ratio about the y-axis is

$$K_y L/r_y = 15 \times 12/1.92$$
$$= 93.75 \ldots \text{governs}$$

From AISC Part 4, Table 4-1, the design axial strength is

$$P = 369 \text{ kips}$$

1.3 c. The shear force along boundary AB is

$$V = pL/2$$
$$= 280 \times 60/2$$
$$= 8400 \text{ lb}$$

The unit shear along boundary AB is

$$q_{AB} = V/B$$
$$= 8400/20$$
$$= 420 \text{ lb/ft}$$

From IBC Table 2306.3.1, a nail spacing of 4 inches at the diaphragm boundary provides an allowable unit shear of

$$q = 425 \text{ lb/ft}$$

1.4 a. The maximum live-load moment, not including impact, at midspan produced by loading one design lane with HS20-44 standard truck loading is obtained from AASHTO Appendix A as

$$M_L = 627.9 \text{ kip-ft}$$

From AASHTO Section 3.8.2.2, the length of loaded span that produces the maximum stress is

$$L = 50 \text{ ft}$$

The impact factor is given by AASHTO Eq. (3-1) as

$$\begin{aligned} I &= 50/(L+125) \\ &= 50/(50+125) \\ &= 0.29 \\ &< 0.30 \ldots \text{satisfactory} \end{aligned}$$

The maximum live-load moment, including impact, is given by

$$\begin{aligned} M_I &= 1.29 M_L \\ &= 1.29 \times 627.9 \\ &= 810 \text{ kip-ft} \end{aligned}$$

For a beam spacing of less than 10 feet, AASHTO Table 3.2.3.1 gives a girder load distribution factor of

$$\begin{aligned} G &= S/6 \\ &= 9/6 \\ &= 1.5 \end{aligned}$$

The maximum live-load moment in an interior beam is

$$\begin{aligned} M &= M_I G/2 \\ &= 8110 \times 1.5/2 \\ &= 608 \text{ kip-ft} \end{aligned}$$

1.5 c. The factor of safety against overturning is given by

$$\begin{aligned} \text{FS} &= W(B-\bar{x})/\left(\gamma_w h^3/6 + \gamma_w hB^2/3\right) \\ &= 54{,}000(25-8.61)/(62.4 \times 22^3/6 + 1373 \times 25^2/3) \\ &= 2.23 \end{aligned}$$

1.6 c. For a site with an undetermined soil profile, the site classification specified by ASCE Section 11.4.2 is D.

From ASCE Table 11.4-1, the site coefficient is

$$F_a = 1.2$$

From ASCE Table 11.4-2, the site coefficient is

$$F_v = 1.5$$

The design response acceleration is given by ASCE Eqs. (11.4-1) and (11.4-3) as

$$S_{DS} = 2F_a S_s/3$$
$$= 0.60$$

The design response acceleration is given by ASCE Eqs. (11.4-2) and (11.4-4) as

$$S_{D1} = 2F_v S_1/3$$
$$= 0.50$$

From ASCE Section 11.4.5,

$$T_S = S_{D1}/S_{DS}$$
$$= 0.83 \text{ second}$$
$$> T_a \ldots \text{ASCE Eq. (12.8-2) applies}$$

For a standard occupancy building, the importance factor is defined in ASCE Tables 1-1 and 11.5-1 as

$$I = 1.0$$

For a special masonry bearing wall structure, the response modification coefficient is given by ASCE Table 12.2-1 as

$$R = 5.0$$

The seismic response coefficient is

$$C_s = S_{DS}/(R/I)$$
$$= 0.12$$

1.7 d. The relevant properties of the column are

$$b = 23.63 \text{ in.}$$
$$h = 20 \text{ ft}$$
$$A_{st} = 6.32 \text{ in.}^2$$
$$A_n = b^2$$
$$= 558 \text{ in.}^2$$
$$\rho = A_s/A_n$$
$$= 0.0113$$
$$F_s = 0.4 f_y$$
$$= 24 \text{ ksi}$$

The radius of gyration of the column is

$$r = 0.289b$$
$$= 6.83 \text{ in.}$$
$$h/r = 35.14 < 99$$

The allowable axial load is given by BCRMS Eq. (2-17) as

$$P_a = (0.25 f'_m A_n + 0.65 A_{st} F_s)[1.0 - (h/140r)^2]$$
$$= (209.3 + 98.6) \times 0.937$$
$$= 288 \text{ kips}$$

1.8 a. The factored load is given by ACI Eq. (9-2) as

$$w_u = 1.2 \times w_D + 1.6 \times w_L$$
$$= 1.2 \times 3 + 1.6 \times 7$$
$$= 14.8 \text{ kips/ft}$$

The critical section for shear, in accordance with ACI Section 11.1.3.1, is a distance d from the support, and the factored shear force at this section is

$$V_u = w_u(\ell/2 - d)$$
$$= 14.8(10/2 - 1.5)$$
$$= 51.8 \text{ kips}$$

1.9 b. Manning's equation for the discharge is

$$Q = \frac{1}{n} A R^{2/3} S_0^{1/2}$$

In this equation, the flow cross-sectional area is $A = 1.8(1.2) + 1.2^2 = 3.6$ m^2, the wetted perimeter is $P = 1.8 + 1.2(1 + \sqrt{5}) = 5.68$ m, and the hydraulic radius is $R = A/P = 0.633$ m. Hence,

$$Q = \frac{1}{0.013}(3.6)(0.633)^{2/3}(0.0025)^{1/2} = 10.2 \text{ m}^3/\text{s}$$

1.10 d. The Hazen-Williams formula is applicable to the turbulent flow of water in conduits of various cross sections:

$$V = 1.318 C_{HW} R^{0.63} S^{0.54}$$

In this equation, V is the average velocity, C_{HW} is the empirical Hazen-Williams roughness coefficient, R is the hydraulic radius, and $S = h_L/L$ is the slope of the energy grade line. For a circular pipe,

$$Q = AV = \frac{\pi}{4} D^2 V$$

Consequently,

$$Q = \left(\frac{\pi}{4} D^2\right) 1.318 C_{HW} \left(\frac{D}{4}\right)^{0.63} \left(\frac{h_L}{L}\right)^{0.54} = 0.432 C_{HW} D^{2.63} \left(\frac{h_L}{L}\right)^{0.54}$$

In analyzing the supply of water from pipe AB to point F, it is computationally convenient to replace pipes BCDF and BEF with a single pipe that has the diameter of pipe AB and otherwise behaves hydraulically like the proposed 6 in. lines. Thus, the discharge in the two 6 in. lines must match the discharge in the new line, and the head loss in each pathway must be the same. Using this information in the previous equation gives

$$Q = 0.432 C_{HW} \left(\frac{8}{12}\right)^{2.63} \left(\frac{h_L}{L_8}\right)^{0.54}$$

$$= 0.432 C_{HW} \left(\frac{6}{12}\right)^{2.63} \left[\left(\frac{h_L}{1800}\right)^{0.54} + \left(\frac{h_L}{1000}\right)^{0.54}\right]$$

$$\left(\frac{1}{L_8}\right)^{0.54} = \left(\frac{6}{8}\right)^{2.63} \left[\left(\frac{1}{1800}\right)^{0.54} + \left(\frac{1}{1000}\right)^{0.54}\right]$$

and $L_8 = 1470$ ft. The equivalent pipe consists of 1470 ft of 8 in. pipe.

1.11 c. The most efficient rectangular section is achieved when $R = y/2$, which is associated with a width $b = 2y$. Substitution of these data into the metric version of the Manning equation gives

$$Q = 22.0 = \frac{1}{0.017}(2y^2)\left(\frac{y}{2}\right)^{2/3} (0.01)^{1/2}$$

which leads to $y^{8/3} = 2.97$ and $y = 1.50$ m. Thus, $b = 2y = 3.0$ m, and the channel depth, including the freeboard, should be 2.0 m.

1.12 c. An instantaneous water balance may be written

$$I - Q = \frac{dS}{dt}$$

in which I and Q are the inflow and outflow rates and S, the storage, is the amount of fluid contained in the water body, relative to some datum. Upon integration over time, the equation may be written as

$$\left(\sum_i P_i - \sum_i E_i\right) A - G = (\text{WSE}_{final} - \text{WSE}_{initial}) A$$

in which P_i, E_i, and the WSEs are entries in the data table and G is the amount of the leak to groundwater for the week. After rearrangement,

$$G = \left(\text{WSE}_{initial} - \text{WSE}_{final} + \sum_i P_i - \sum_i E_i\right) A$$

or

$$G = \left[250.00 - 249.25 + \frac{1.15 - 2.40}{12}\right](75)(60)$$

Thus, $G = 2900$ ft^3/week $= (2900$ ft^3/week$)(7.48$ gallons/ft$^3)$ (1 week/7 days) $= 3100$ gal/day.

1.13 d. The rational method will be applied. The correct intensity factor is $i = 1.2$ in./h. To maximize runoff, the highest reasonable runoff coefficient should be selected from Table 6-1 in *Civil Engineering PE License Review*; for a single-family residential area, pick $C = 0.50$. The method then gives

$$Q_p = CiA = (0.5)(1.2)(60) = 36 \text{ acre-in./h} \approx 36 \text{ ft}^3/\text{s}$$

1.14 c. Darcy's law is $V = -Ki$. The discharge Q may be written $Q = VA$; in this case, the area $A = bw$ with the width w being 1 kilometer. The transmissibility T is the product of the aquifer width b and the hydraulic conductivity K. Assembling these relations,

$$Q = VA = (-i)Kbw = \left(\frac{0.25 \text{ m}}{1000 \text{ m}}\right)\left(40 \frac{\text{m}^2}{\text{day}}\right)(1000 \text{ m}) = 10 \frac{\text{m}^2}{\text{day}}$$

1.15 b. Convert all cations and anions to common units of calcium carbonate (50 mg/L $CaCO_3$ meq).

	Concentration (mg/L)	MW/Charge (mg/meq)	Concentration (meq/L)	Concentration (meq/L as $CaCO_3$)
Ca^{2+}	71	20	3.55	177
Mg^{2+}	19	12.2	1.56	78
Na^+	11	23	0.48	27
HCO_3^-	120	61	1.97	98
SO_4^{2-}	101	48	2.10	105
Cl^-	55	35.5	1.55	78

$$\sum \text{cations} = [Ca^{2+}] + [Mg^{2+}] + [Na^+]$$
$$= 177 \text{ mg/L} + 78 \text{ mg/L} + 24 \text{ mg/L}$$
$$= 279 \text{ mg/L}$$

$$\sum \text{anions} = [HCO_3^-] + [SO_4^{2-}] + [Cl^-]$$
$$= 98 \text{ mg/L} + 105 \text{ mg/L} + 78 \text{ mg/L}$$
$$= 281 \text{ mg/L}$$

1.16 d. Find DO after mixing in the river:

$$C_0 = \frac{Q_r C_r + Q_w C_w}{Q_r + Q_w}$$

$$= \frac{(1 \text{ m}^3/\text{s})(8 \text{ mg/L}) + (0.1 \text{ m}^3/\text{s})(1.5 \text{ mg/L})}{(1 \text{ m}^3/\text{s} + 0.1 \text{ m}^3/\text{s})}$$

$$= 7.41 \text{ mg/L}$$

$$DO_{sat} @ 20°C = 9.17 \text{ mg/L}$$
$$D_0 = DO_{sat} - DO$$
$$= 9.17 \text{ mg/L} - 7.41 \text{ mg/L}$$
$$= 1.76 \text{ mg/L}$$

DO_{sat} depends on the temperature of the waste and the river.

1.17 b.

$$\text{SF} = \frac{(\Delta \text{TSS})(Q)}{A}$$

$$= \frac{(300 \text{ mg/L})(0.7)\left(3 \times 10^6 \frac{\text{gal}}{\text{d}}\right)\left(\frac{3.78 \text{ L}}{\text{gal}}\right)\left(\frac{1 \text{ lb}}{453 \times 10^3/\text{mg}}\right)}{(60 \text{ ft})^2(3.14)/4}$$

$$= 1.86 \text{ lb/ft}^2\text{-d}$$

1.18 c.

$$\text{TSS removed} = (0.70)(300 \text{ mg/L})\frac{(3 \times 10^6 \text{ gal})}{\text{d}}\left(\frac{3.78 \text{ L}}{\text{gal}}\right)$$

$$= \left(\frac{1 \text{ kg}}{10^6 \text{ mg}}\right)$$

$$= 2380 \text{ kg/d}$$

1.19 c.

$$Q_c = \frac{X_{\text{TSS}} V}{M}$$

$$M_{\text{wasted}} = \frac{X_{\text{TSS}} V}{\theta_c} = \left(\frac{(3000 \text{ mg/L})(1.5 \times 10^6/\text{L})}{5 \text{d}}\right) = \left(\frac{1 \text{ kg}}{10^6/\text{mg}}\right)$$

$$= 900 \text{ kg/d}$$

1.20 b.

$$YQ(S_0 - S) = M_{\text{wasted}}$$

$$Y_{\text{obs}} = \frac{M_{\text{wasted}}}{Q(S_0 - S)} = \frac{(900 \text{ kg TSS/d})\left(\frac{3500 \text{ mg TUSS}}{3000 \text{ mg TSS}}\right)\left(\frac{10^6 \text{ mg}}{1 \text{ kg}}\right)}{(7.56 \times 10^6 \text{ L/d})(200 \text{ mg/L} - 10 \text{ gm/L})}$$

$$= 0.52 \text{ mg TUSS/mg BOD removed}$$

1.21 d.

1.22 b.

$$C = C_0 e^{-k_1 \cdot t_v} = C_0^{-k_1 \frac{V}{Q}}$$

$$V = -\frac{Q}{K} \ln \frac{C}{C_0} = -\frac{(12{,}000 \text{ L/h})}{(0.35/\text{h})} \ln\left(\frac{1}{10 \times 10^6}\right)$$

$$= 5.5 \times 10^5/\text{L}$$

1.23 c. Assume $S_f = 2.5$, $S_v = 1.0$. For solids in primary sludge, (Eq. 8.41) in *Civil Engineering PE License Review* gives

$$\frac{1}{S_s} = \frac{0.33}{2.5} + \frac{0.67}{1.0} = 0.80$$

$$S_s = 1.25$$

For the primary sludge, Eq. (8.41) gives

$$\frac{1}{S_{sl}} = \frac{0.05}{1.25} + \frac{0.95}{1} = 0.99$$

$$S_{sl} = 1.01$$

1.24 c. For solids after digestion,

$$\%V_s = \frac{(0.50)(670 \text{ lb})}{(330 \text{ lb}) + (0.50)(670 \text{ lb})} \times 100 = 50\%$$

$$\frac{1}{S_s} = \frac{0.50}{2.5} + \frac{0.50}{1.0} = 0.70$$

$$S_s = 1.43$$

1.25 c.

$$e = \frac{V_v}{V_s}$$

$$\text{Weight of dry soil} = W_s = \frac{\text{weight of moist soil}}{1 + \frac{w\%}{100}} = \frac{2.63}{1 + \frac{11}{100}} = 2.369 \text{ kN}$$

$$\text{Volume of soil solids} = V_s = \frac{W_s}{G_s \gamma_w} = \frac{2.369}{(2.67)(9.81)} = 0.0904 \text{ m}^3$$

$$\text{Volume of voids} = V_v = V - V_s = 0.15 - 0.0904 = 0.0596 \text{ m}^3$$

$$e = \frac{0.0596}{0.0904} = 0.66$$

1.26 b.

$$\tan\phi = \frac{S}{N} = \frac{28}{40}; \quad \phi = 35°$$

1.27 b.

$$\text{Relative density, } D_r = \frac{e_{max} - e}{e_{max} - e_{min}}$$

$$0.459 = \frac{0.82 - e}{0.82 - 0.45}$$

$$e = 0.65$$

$$\gamma_d = \frac{G_s \gamma_w}{1 + e} = \frac{(2.66)(62.4)}{1 + 0.65} = 100.6 \text{ lb/ft}^3$$

1.28 b. $\gamma_d = \dfrac{\gamma}{1+\dfrac{w}{100}}$ The following table can now be prepared.

Moisture Content, w, %	Moist Unit Weight, γ, lb/ft³	Dry Unit Weight, γ_d, lb/ft³
12	113.7	101.5
14	123.1	108
16	128.2	110.5
18	131.6	111.5
20	131.4	109.5
22	130.5	107

The plot of γ_d versus w is shown in Exhibit 1.28. From this, the maximum dry unit weight is about 112 lb/ft³.

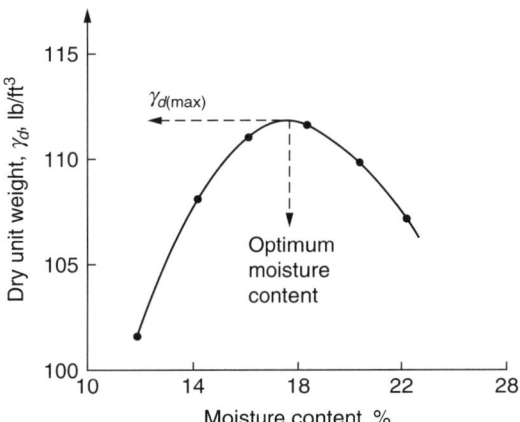

Exhibit 1.28

1.29 b.

$$\sigma'_B = \sigma_B - u_B$$
$$\sigma_B = (2\text{ m})(14.6\text{ kN/m}^3) + (1\text{ m})(17.8\text{ kN/m}^3) = 47\text{ kN/m}^2$$
$$u_B = (1\text{ m})(\gamma_w) = (1\text{ m})(9.81\text{ kN/m}^3) = 9.81\text{ kN/m}^2$$
$$\sigma'_B = 47 - 9.81 = 37.19\text{ kN/m}^2$$

1.30 d.

$$q = k\dfrac{N_f}{N_d}H$$

N_f = Number of flow channels
N_d = Number of drops

$$q = (2.5 \times 10^{-5})\left(\dfrac{8}{12}\right)(20-5) = 25 \times 10^{-5}\text{ ft}^3/\text{min/ft}$$

1.31 c.

Given $c = 0$.

$$q_u = qF_{qs}F_{qd}N_q + \frac{1}{2}\gamma BF_{\gamma s}F_{\gamma d}N_\gamma$$
$$q = (\gamma)(1 \text{ m}) = (16)(1) = 16 \text{ kN/m}^2$$

For $\phi = 35°$, $N_q = 33.3$ and $N_\gamma = 48.03$ (see Fig. 9.18 in *Civil Engineering PE License Review*).

$$F_{qs} = 1 + \left(\frac{B}{L}\right)\tan\phi = 1 + \left(\frac{1.5}{1.5}\right)\tan 35° = 1.7$$

$$F_{\gamma s} = 1 - 0.4\left(\frac{B}{L}\right) = 1 - 0.4\left(\frac{1.5}{1.5}\right) = 0.6$$

$$F_{qd} = 1 + 2\tan\phi(1-\sin\phi)^2\left(\frac{D_f}{B}\right) = 1 + 2\tan 35°(1-\sin 35°)^2\left(\frac{1}{1.5}\right) = 1.24$$

$$F_{\gamma d} = 1$$

$$q_u = (16)(1.7)(1.24)(33.3) + \tfrac{1}{2}(16)(1.5)(48.03)(0.6)(1) = 1469 \text{ kN/m}^2$$

1.32 b.

For triaxial tests:

$$\sigma_1 = \sigma_3 \tan^2\left(45° + \frac{\phi}{2}\right) + 2c\tan\left(45° + \frac{\phi}{2}\right)$$

$$69 = 19\tan^2\left(45° + \frac{\phi}{2}\right) + 2c\tan\left(45° + \frac{\phi}{2}\right)$$

$$91 = 25\tan^2\left(45° + \frac{\phi}{2}\right) + 2c\tan\left(45° + \frac{\phi}{2}\right)$$

From the two preceding equations,

$$22 = 6\tan^2\left(45° + \frac{\phi}{2}\right)$$

$$\phi = 2\left[\tan^{-1}\left(\frac{22}{6}\right)^{1/2} - 45°\right] = 34.85°$$

Back-substituting $\phi = 34.85°$, we obtain $c \approx 0$. So the soil is cohesionless.

Rankine's active force:

$$P_a = \frac{1}{2}\gamma H^2 K_a$$

$$K_a = \tan^2\left(45° - \frac{\phi}{2}\right) = \tan^2\left(45° - \frac{34.85°}{2}\right) = 0.273$$

$$\gamma = \frac{G_s\gamma_w(1+w)}{1+e} = \frac{(2.65)(62.4)(1+0.1)}{1+0.5} = 121.26 \text{ lb/ft}^3$$

$$P_a = \frac{1}{2}(121.26)(20^2)(0.273) = 6621 \text{ lb/ft}$$

1.33 c.

$$v_p = 2700/[(2)(0.90)(0.87)(0.92)] = 1874 \text{ pc/h/lane}$$

From Exhibit 23-2 (TRB, 2000), for a free-flow speed of 70 mph, the maximum service flow rate for level of service C is 1770 pc/h/lane, and the maximum service flow rate for level of service D is 2150 pc/h/lane.

1770 < 1874 < 2150, so the level of service is D.

1.34 c. Level of service is defined in terms of control delay. Control delay is given by

$$d = d_1 \text{ PF} + d_2 + d_3$$

$$d_1 = \frac{0.5C(1 - g/C)^2}{1 - [\min(1, X)](g/C)}$$

$$g/C = 22 \text{ s}/60 \text{ s} = 0.367$$

$$d_1 = 0.50(60)(1 - 0.367)2/[1 - (0.85)(0.367)] = 17.5 \text{ s}$$

Since the intersection is isolated, PF = 1.0.

$$d_2 = 900T\{(X-1) + [(X-1)^2 + (8kIX/cT)]^{1/2}\}$$

Since the intersection is isolated, $I = 1.0$. Since the signal is pretimed, $k = 0.5$.

$$c = \text{capacity} = (g/C)s = (0.367)(3300) = 1211 \text{ veh/h}$$
$$d_2 = 900(0.25)\{(0.85 - 1) + [(0.85 - 1)^2$$
$$+ (8 \times 0.5 \times 1 \times 0.85)/(0.25 \times 1211)]^{1/2}\} = 7.5 \text{ s}$$
$$d_3 = 0 \text{ since there is no initial queue}$$
$$d = (17.5)(1) + 7.5 + 0 = 25.0 \text{ s}$$

From Exhibit 16-2 of the HCM, control delay per vehicle of 25.0 s corresponds to level of service C.

1.35 b. Between stations 10 + 00 and 10 + 40, the cut volume is calculated by the average end area method and the fill volume as a pyramid.

$$V_c = \frac{(25 + 15)(40)}{(2)(27)} = 29.6 \text{ yd}^3$$

$$V_f = \frac{(12)(55)}{(3)(27)} = 5.9 \text{ yd}^3$$

Between stations 10 + 40 and 10 + 95, fill is calculated by the average end area method and cut as a pyamid.

$$V_c = \frac{(15)(55)}{(3)(27)} = 10.2 \text{ yd}^3$$

$$V_f = \frac{(12 + 20)(55)}{(2)(27)} = 32.6 \text{ yd}^3$$

Total cut = 29.6 + 10.2 = 39.8 yd³, and total fill = 5.9 + 32.6 = 38.5 yd³. Accounting for shrinkage, the cut will provide (39.8)(1.00 − 0.20) = 31.8 yd³ of fill. Therefore, material will have to be borrowed. Borrow = 38.5 − 31.8 = 6.7 yd³ of compacted fill.

1.36 d.

$$d_s = d_p + d_b = 0.278vt + \frac{V^2}{254\left(\frac{a}{9.81} \pm G\right)}$$

The AASHTO Green Book (AASHTO, 2004) recommends that 2.5 s be used for the reaction time t and 3.4 m/s² as the deceleration rate for stopping sight distance.

$$d_s = 0.278(70)(2.5) + \frac{70^2}{254\left(\frac{3.4}{9.81} + 0.03\right)} = 48.65 + 51.23 = 99.88 \text{ m}$$

1.37 d. The semitangent distance T is

$$T = R \tan(\Delta/2) = 1200 \tan(28°/2) = 299 \text{ ft}$$

The station of the PC is $(5 + 43) + (13 + 22) - (2 + 99) = 15 + 66$.

The length of the curve is

$$L = R\Delta(\pi/180°) = 586 \text{ ft}$$

The station of the PT is $(15 + 66) + (5 + 86) = 21 + 52$.

1.38 c. The middle ordinate M is measured from the centerline of the inside lane. Therefore,

$$M = \frac{1}{2}(3.6) + 8.2 = 10.0 \text{ m}$$

The radius of the centerline of the inside lane is

$$R = 400 - \frac{1}{2}(3.6) = 398.2 \text{ m}$$

From Exhibit 3-53 of the AASHTO Green Book (AASHTO, 2004, p. 226), the stopping sight distance S is approximately 180 m.

1.39 b. The algebraic difference in grade A is

$$A = g_1 - g_2 = 2.5 - (-3.5) = 6\%$$

From Exhibit 3-71 in the AASHTO Green Book (AASHTO, 2004, p. 271), the minimum length of the vertical curve is approximately 680 ft.

1.40 b. The distance from the PVC to the low point is given by

$$x = -g_1 K$$
$$A = g_2 - g_1 = 3.6 - (-2.3) = 5.9\%$$
$$K = L/A = 300/5.9 = 50.8 \text{ m}$$
$$x = -(-2.3)(50.8) = 116.9 \text{ m}$$

Solutions to Chapter 2 Problems **97**

Station of the low point = $(7 + 43) + (1 + 16.9) = 8 + 59.9$

Elevation of the low point = $E_x = E_{PVC} + g_1 x + (A/2L)x^2 = 87.621$
$+ (-2.3)(1.169) + [(5.9/3)(1.169)^2]/2 = 86.276$ m

SOLUTIONS TO CHAPTER 2 PROBLEMS

2.1 **a.** TSS_{eff} affects sludge wastage [Eq. (8.34) in *Civil Engineering PE License Review*].

2.2 **c.**

$$UR = \frac{Q_r}{Q} \text{ OR}$$
$$= (0.5)(600 \text{ gpd/ft}^2)$$
$$= 300/\text{gpd/ft}^2$$

2.3 **c.**

$$V_r = \frac{300 \text{ gal}}{\text{ft}^2\text{-d}} \times \frac{1 \text{ ft}^3}{7.48 \text{ gal}} = \frac{40 \text{ ft}}{\text{d}} = \frac{1.67 \text{ ft}}{\text{h}}$$
$$V = V_i + V_r$$
$$= 3.8 \text{ ft/h} + 1.67 \text{ ft/h} = 5.47 \text{ ft/h}$$

2.4 **a.**

$$Q = \left(\frac{1 \times 10^6 \text{ gal}}{\text{d}}\right)\left(\frac{1 \text{ m}^3}{3780 \text{ L}}\right)$$
$$= 260 \text{ m}^3/\text{d}$$
$$A = \frac{Q}{HLR}$$
$$= \frac{260 \text{ m}^3/\text{d}}{50 \text{ m}^3/\text{m}^2\text{-d}}$$
$$= 5.2 \text{ m}^2$$

2.5 **d.**

$$\text{BOD loading} = C_{BOD} Q$$
$$= (0.67)(300 \text{ mg/L})(263 \text{ m}^3/\text{d})\left(\frac{1 \text{ kg}}{106 \text{ mg}}\right)\left(\frac{1000 \text{ L}}{1 \text{ m}^3}\right)$$
$$= 53 \text{ kg/d}$$
$$A = \frac{\text{BOD loading}}{\text{BOD loading rate}}$$
$$= \frac{53 \text{ kg/d}}{2 \text{ kg/m}^2\text{-d}}$$
$$= 26 \text{ m}^2$$

2.6 b. From Fig. 8.9 in *Civil Engineering PE License Review*, for BOD = 200 mg/L, the HLR must be less then 0.2 m³/m²-d.

2.7 a.

2.8 c.

$$Al_2(SO_4)_3 \cdot 18H_2O$$
$$MW = 666$$

$$Alum = \frac{10 \text{ mg P}}{L} \times \frac{1 \text{ mol/P}}{31 \text{ g P}} \times \frac{\frac{2.3}{2} \text{ mol alum}}{1 \text{ mol P}} \times \frac{666 \text{ g}}{1 \text{ mol alum}}$$
$$= 354 \text{ mg/L}$$

2.9 b.

$$\theta_c = \theta_h = 15 \text{ d}$$
$$V = (\theta_c)(Q)$$
$$= (15 \text{ d})(50 \text{ m}^3/\text{d}) = 750 \text{ m}^3$$

2.10 d.

2.11 b.

$$\Delta X_{cells} = \frac{Y \Delta BOD_c}{1 + K_d \theta_c}$$
$$= \frac{(0.075)(160 \text{ kg/d})}{1 + (0.03)(20)}$$
$$= 11 \text{ kg/d}$$

2.12 b.

$$V_M = \left(\frac{5.6 \text{ ft}^3}{\text{lb BOD}}\right)\left(\frac{2.2 \text{ lb BOD}}{1 \text{ kg BOD}}\right)\left(\frac{160 \text{ kg}}{\text{d}} - \frac{11 \text{ kg}}{\text{d}}\right)$$
$$= 1835 \text{ ft}^3/\text{d}$$

2.13 c. To meet the variable oxygen demand.

2.14 c. Using the average daily flow and the peaking factor given in the problem statement, peak flow is calculated as

$$Q_{peak} = 3Q_{ave} = (3)(5.5 \text{ MGD})\left(1.547 \frac{\text{cfs}}{\text{MGD}}\right) = 25.53 \text{ cfs}$$

The AGC tank volume can now be calculated from the design detention time:

$$t_R = \frac{V}{Q} \Rightarrow V = t_R \times Q = (1 \text{ min})\left(60 \frac{\text{s}}{\text{min}}\right)\left(25.53 \frac{\text{ft}^3}{\text{s}}\right) = 1532 \text{ ft}^3$$

Using the dimension ratios we know that

$$L = W = 1.5D$$

Therefore,

$$V = LWD = (1.5D)(1.5D)D = 2.25D^3 = 1532 \text{ ft}^3$$

Rearranging and solving for D gives us L and W:

$$D = \left(\frac{1532 \text{ ft}^3}{2.25}\right)^{1/3} = 8.8 \text{ ft}$$

$$L = W = 1.5D = (1.5)(8.8 \text{ ft}) = 13.2 \text{ ft}$$

2.15 b. Using the design detention time to find clarifier volume as in Problem 2.14, we get

$$V = t_R \times Q = (2 \text{ h})\left(\frac{1 \text{ day}}{24 \text{ h}}\right)\left(5.5 \times 10^6 \frac{\text{gal}}{\text{day}}\right)\left(\frac{1 \text{ ft}^3}{7.48 \text{ gal}}\right) = 61{,}275 \text{ ft}^3$$

Using the expression for overflow rate, we can find tank surface area:

$$V_O = \frac{Q}{A_S} \quad \Rightarrow \quad A_S = \frac{Q}{V_O} = \frac{5.5 \times 10^6 \text{ gpd}}{1000 \frac{\text{gpd}}{\text{ft}^2}} = 5500 \text{ ft}^2$$

Now tank depth can be determined:

$$\text{depth} = \frac{V}{A_S} = \frac{61{,}275 \text{ ft}^3}{5500 \text{ ft}^2} = 11.14 \text{ ft}$$

2.16 a. The critical (maximum) deficit occurs at the minimum DO concentration:

$$D_c = \text{DO}_{\text{sat}} - \text{DO}_{\text{min}}$$

The determination of DO_{sat} requires knowledge of the temperature of the mixed flows, and D_c could be calculated using the well-known Streeter-Phelps equation as follows:

$$D_c = \frac{k_d L_0}{k_a - k_d}[\exp(-k_d t_c) - \exp(-k_a t_c)] + D_0 \exp(-k_a t_c)$$

where

$$t_c = \frac{1}{k_a - k_d} \ln\left[\frac{k_a}{k_d}\left(1 - D_0 \frac{k_a - k_d}{k_d L_0}\right)\right]$$

First, check the biological use coefficient (k_d) and the reaeration coefficient (k_a) at the temperature of the mixed flows (T_m). If the use coefficient is larger than the reaeration coefficient, oxygen is being consumed faster than it is being replenished, causing a deficit. To calculate the flow-weighted temperature, the flow rates of the river and wastewater must be determined in the same units as follows:

$$Q_r = (40 \text{ ft})(3 \text{ ft})\left(1\frac{\text{ft}}{\text{s}}\right) = 120 \frac{\text{ft}^3}{\text{s}}$$

and

$$Q_{ww} = (5 \text{ MGD})\left(1.547 \frac{\text{cfs}}{\text{MGD}}\right) = 7.735 \frac{\text{ft}^3}{\text{s}}$$

T_m may now be calculated:

$$T_m = \frac{Q_{ww}T_{ww} + Q_rT_r}{Q_{ww} + Q_r} = \frac{(7.735 \text{ cfs})(16°C) + (120 \text{ cfs})(22°C)}{7.735 \text{ cfs} + 120 \text{ cfs}} = 21.64°C$$

Correction can now be made to k_d and k_a through the temperature correction expression as follows:

$$k_T = k_{20}\, \theta^{(T-20)}$$

$$(k_d)_{21.64} = (k_d)_{20}\, 1.056^{(21.64-20)} = (0.3)\, 1.056^{1.64} = 0.328 \text{ day}^{-1}$$

$$(k_a)_{21.64} = (k_a)_{20}\, 1.024^{(21.64-20)} = (0.25)\, 1.024^{1.64} = 0.260 \text{ day}^{-1}$$

Since k_d is greater than k_a, a deficit will occur. To calculate the deficit, the BOD concentration of the mixed flows is required and may be determined as follows:

$$L_o = \frac{Q_{ww}L_{ww} + Q_rL_r}{Q_{ww} + Q_r} = \frac{(7.735 \text{ cfs})\left(20\, \frac{\text{mg}}{\text{L}}\right) + (120 \text{ cfs})\left(0.9\, \frac{\text{mg}}{\text{L}}\right)}{7.735 \text{ cfs} + 120 \text{ cfs}}$$

$$= 2.057\, \frac{\text{mg}}{\text{L}}$$

The initial deficit (D_0) is also needed, which requires the initial DO in the river. This may be determined by finding 90 percent of the saturation value at the river temperature as follows:

$$(\text{DO}_i)_r = (0.9)(\text{DO}_{\text{sat}})_{22°C} = (0.9)\left(8.8\, \frac{\text{mg}}{\text{L}}\right) = 7.92\, \frac{\text{mg}}{\text{L}}$$

This value can be used to determine the flow-weighted DO as follows:

$$\text{DO}_0 = \frac{Q_{ww}\text{DO}_{ww} + Q_r\text{DO}_r}{Q_{ww} + Q_r} = \frac{(7.735 \text{ cfs})\left(1\, \frac{\text{mg}}{\text{L}}\right) + (120 \text{ cfs})\left(7.92\, \frac{\text{mg}}{\text{L}}\right)}{7.735 \text{ cfs} + 120 \text{ cfs}}$$

$$= 7.501\, \frac{\text{mg}}{\text{L}}$$

From this and the value of DO_{sat} at T_m, the initial deficit may be determined:

$$D_0 = (DO_{sat})_{T_m} - DO_0 = (DO_{sat})_{21.64°C} - 7.50 \frac{mg}{L}$$

$$= 8.865 - 7.501 = 1.364 \frac{mg}{L}$$

Plugging these values into the equation for critical time (t_c) above yields:

$$t_c = \frac{1}{0.26 - 0.328} \ln\left[\frac{0.26}{0.328}\left(1 - (1.364)\frac{0.26 - 0.328}{(0.328)(2.057)}\right)\right] = 1.522 \text{ days}$$

The units are not shown in the previous expression for simplicity, and it is left to the examinee to verify the homogeneity of the units. This value must be put into the deficit equation as follows:

$$D_c = \frac{(0.328)(2.057)}{0.26 - 0.328}[\exp(-0.328 \times 1.522) - \exp(-0.26 \times 1.522)]$$
$$+ 1.364 \exp(-0.26 \times 1.522)$$
$$D_c = 1.575 \frac{mg}{L}$$

Again, verification of the homogeneity of the units is left to the examinee. Finally, DO_{min} may be calculated as follows:

$$DO_{min} = (DO_{sat})_{T_m} - D_c = 8.865 - 1.575 = 7.29 \frac{mg}{L}$$

A problem of this complexity would never be given as a single question on the Principles and Practice exam, where time is of the essence. However, it is a useful exercise in that it emphasizes the importance of understanding all of the steps involved in using the Streeter-Phelps equation, which should benefit the examinee.

2.17 b. As with Problem 2.16, this problem starts by calculating the temperature-corrected constants k_d and k_a as follows:

$$Q_r = (40 \text{ ft})(4 \text{ ft})\left(1\frac{\text{ft}}{\text{s}}\right) = 160 \frac{\text{ft}^3}{\text{s}}$$

and

$$Q_{ww} = (5 \text{ MGD})\left(1.547 \frac{\text{cfs}}{\text{MGD}}\right) = 7.735 \frac{\text{ft}^3}{\text{s}}$$

$$T_m = \frac{Q_{ww}T_{ww} + Q_r T_r}{Q_{ww} + Q_r} = \frac{(7.735 \text{ cfs})(13°C) + (160 \text{ cfs})(4°C)}{7.735 \text{ cfs} + 160 \text{ cfs}} = 4.415°C$$

$$(k_d)_{4.415} = (k_d)_{20}\, 1.056^{(4.415-20)} = (0.3)\, 1.056^{-15.585} = 0.128\ \text{day}^{-1}$$

$$(k_a)_{4.415} = (k_a)_{20}\, 1.024^{(4.415-20)} = (0.25)\, 1.024^{-15.585} = 0.173\ \text{day}^{-1}$$

Since the reaeration constant is greater than the biological use constant, there will never be a demand for oxygen that depletes the river. Therefore, the initial flow-weighted DO will be the minimum value. This may be calculated as in Problem 2.16; the initial DO in the river is determined by finding 90 percent of the saturation value at the river temperature as follows:

$$(DO_i)_r = (0.9)(DO_{sat})_{4°C} = (0.9)\left(13.1\ \frac{\text{mg}}{\text{L}}\right) = 11.79\ \frac{\text{mg}}{\text{L}}$$

This value can be used to determine the flow-weighted DO as follows:

$$DO_o = \frac{Q_{ww} DO_{ww} + Q_r DO_r}{Q_{ww} + Q_r} = \frac{(7.735\ \text{cfs})\left(1\ \frac{\text{mg}}{\text{L}}\right) + (160\ \text{cfs})\left(11.79\ \frac{\text{mg}}{\text{L}}\right)}{7.735\ \text{cfs} + 160\ \text{cfs}}$$

$$= 11.29\ \frac{\text{mg}}{\text{L}}$$

2.18 c. Cyanobacteria are aerobic autotrophs and use H_2O as their primary photosynthetic electron donor. Purple and green bacteria share anaerobic metabolic processes, photosynthetic electron donors, and both heterotrophic and autotrophic carbon sources.

2.19 a. The hydraulic conductivity of an unconfined aquifer can be calculated from

$$K = \frac{Q \ln\left(\frac{r_2}{r_1}\right)}{\pi\left(h_2^2 - h_1^2\right)}$$

Values for Q, r_1, and r_2 were given in the problem statement, along with the drawdown in each well. Water depth in each well is determined by subtracting the drawdown from the aquifer thickness. Aquifer thickness can be found by taking the difference between the depth to the confining layer and the depth to the groundwater table, which is 40 ft – 12 ft, or 28 ft thick. Water depth in each well can now be determined as follows:

$$h_1 = 28 - 4.6 = 23.4\ \text{ft} \quad \text{and} \quad h_2 = 28 - 0.3 = 27.7\ \text{ft}$$

Therefore, the hydraulic conductivity of the aquifer can be determined as follows:

$$K = \frac{(5 \text{ gpm}) \ln\left(\frac{40}{30}\right)}{\pi[(27.7 \text{ ft})^2 - (23.4 \text{ ft})^2]} = 0.002084 \frac{\text{gpm}}{\text{ft}^2} \times \frac{1 \text{ ft}^3}{7.48 \text{ gal}}$$
$$= 0.000279 \frac{\text{ft}}{\text{min}}$$

$$\left(0.000279 \frac{\text{ft}}{\text{min}}\right)\left(60 \frac{\text{min}}{\text{h}}\right)\left(24 \frac{\text{h}}{\text{day}}\right)\left(365 \frac{\text{day}}{\text{yr}}\right) = 146.4 \frac{\text{ft}}{\text{yr}}$$

2.20 a. If the sedimentation tank is sized to remove 100 percent of the largest particles, this is the same as saying that the critical settling velocity (V_{sc}) in the tank corresponds to the settling velocity of the largest particle. The removal fraction R of the smallest particles is given by $R = V_s/V_{sc}$, where V_s is the settling velocity of the smallest particles.

The settling velocity of a particle settling according to Type I settling is given by Stokes's Law:

$$V_s = \frac{g(\rho_p - \rho_w)d_p^2}{18\mu}$$

where
   V_s = settling velocity (m/s)
   g = acceleration due to gravity (9.81 m/s^2)
   ρ_p = particle density (1.8 × 1000 kg/m^3 = 1800 kg/m^3)
   ρ_w = water density (1000 kg/m^3)
   d_p = particle diameter (10.10^{-6} m for the smallest flocs and 85 × 10^{-6} m for the largest flocs)
   μ = viscosity of water (0.001 kg/m.s)

For this problem, $V_{sc} = V_s$ for the largest particle. V_s can be found from Stokes's Law, given $d_p = 8.5 \times 10^{-6}$ m, as 3.1×10^{-3} m/s. V_s for the smallest particles is also found from Stokes's Law, given $d_p = 10^{-5}$ m, and equals 4.4×10^{-5} m/s. So,

$$R = (4.4 \times 10^{-5})/(3.1 \times 10^{-3}) = 0.014 = 1.4\%$$

2.21 d. This problem appears at first glance to be a pipe-flow problem, but the data do not allow a direct solution via either the Darcy-Weisbach or Hazen-Williams formulations, since it is not known whether the conduit is circular. The Manning equation comes to the rescue, however, as it is routinely applied to the flow of water through various cross sections, and the data here are appropriate for the use of this equation. Since the smallest head loss is sought, Table 5.4 in *Civil Engineering PE License Review* is consulted for the smallest Manning n that is applicable to concrete; one finds $n = 0.01$ for cement with a neat surface as the lowest value. Hence,

$$V = \frac{Q}{A} = \frac{1}{n} R^{2/3} \left(\frac{h_L}{L}\right)^{1/2}$$

or

$$h_L = L\left[\frac{Vn}{R^{2/3}}\right]^2 = (1000)\left[\frac{2.50(0.010)}{(0.24)^{2/3}}\right]^2 = 4.2 \text{ m}$$

since the hydraulic radius is $R = A/P = 0.90/3.75 = 0.24$ m.

2.22 a. One can write an energy-line expression between the two reservoirs, following Example 5.3 of *Civil Engineering PE License Review*:

$$200 - K_{ent}\frac{V_{12}^2}{2g} - \left[f\frac{L}{D}\right]_{12}\frac{V_{12}^2}{2g} - K_{contr}\frac{V_6^2}{2g} - \left[f\frac{L}{D}\right]_6\frac{V_6^2}{2g} - K_{exit}\frac{V_6^2}{2g} = 130$$

Before going further, it is worthwhile to examine the approximate size of the loss coefficients K relative to the size to the products fL/D that are multiplied by the same velocity heads. For the 12 in. line, $fL/D = (0.02)(1000)/1.0 = 20$, but K_{ent} is probably no larger than 0.5, assuming a sharp-edged pipe entrance. For the 6 in. line, $fL/D = 40$, but neither of the Ks is larger than 1.0. Also, by continuity $V_6 = 4V_{12}$, so one can conclude that the omission of the local losses entirely should change the discharge result by no more than about 2 percent. Thus, all of the local loss terms can be dropped.

Using continuity again,

$$V_{12} = \frac{Q}{A_{12}} = \frac{Q}{\frac{\pi}{4}(1.0)^2} = 1.273Q$$

$$V_6 = \frac{Q}{A_6} = \frac{Q}{\frac{\pi}{4}(0.5)^2} = 5.09Q$$

and the energy equation becomes

$$200 - 130 = 70 = \frac{0.02}{2(32.2)}\left[\frac{1000}{1.0}(1.273)^2 + \frac{1000}{0.5}(5.09)^2\right]Q^2$$

with the result $Q = 2.05$ ft³/s. Additional computation would show that the inclusion of all three local loss terms would reduce Q to about 2.02 ft³/s, so the omissions were justified.

2.23 b. In the solution of this problem, it is important to know the discharge per unit channel width, which is $q = Q/10 = 500/10 = 50$ ft²/s. The upstream specific energy at the normal depth of flow is

$$E_1 = y_1 + \frac{V_1^2}{2g} = 3.00 + \frac{1}{2(32.2)}\left(\frac{50}{3}\right)^2 = 3.00 + 4.31 = 7.31 \text{ ft}$$

The critical depth associated with this discharge is

$$y_c = \left(\frac{q^2}{g}\right)^{1/3} = \left(\frac{50^2}{32.2}\right)^{1/3} = 4.27 \text{ ft}$$

The minimum specific energy required atop the rise to pass the given discharge is the critical specific energy E_c, which is

$$E_c = \frac{3}{2}y_c = 6.40 \text{ ft}$$

The actual energy atop the rise is $E_2 = E_1 - 0.5 = 6.81$ ft, which exceeds the minimum required. Hence, the flow, which is initially supercritical, will remain so atop the rise, and one thus expects the depth atop the rise to remain less than the critical depth:

$$E_2 = 6.81 = y_2 + \frac{q^2/2g}{y_2^2} = y_2 + \frac{(50)^2/2(32.2)}{y_2^2} = y_2 + \frac{38.8}{y_2^2}$$

Solve by successive trials:

Trial	y	38.8/y²	$E_2 = 6.81$ ft
1	4.0	2.42	6.42 (low)
2	3.5	3.17	6.67 (low)
3	3.3	3.56	6.86 (a bit high)
4	3.35	3.46	6.81 (OK)

2.24 b. The critical depth for this flow is $y_c = [q^2/g]^{1/3}$ with $q = 25.0/5.00 = 5.00$ m²/s. Thus, the critical depth is $y_c = [5.00^2/9.81]^{1/3} = 1.37$ m. Now turn to Table 5.5 in *Civil Engineering PE License Review*, which displays the various gradually varied flow profiles that are possible. Since the critical depth lies between the initial and final depths, the profile must cross this depth, yet the table shows that no gradually varied flow profile can accomplish this. The only other mechanism for crossing the critical depth is the hydraulic jump (so answer **(d)** can now be ruled out).

Furthermore, if the varied flow were to occur downstream from the break in slope, where the normal depth indicates a subcritical flow and Table 5.5 shows that only *M* profiles are possible, one sees that the only way to reach normal depth is to jump directly to that depth from an initially unknown depth. If instead the flow jumps to a depth other than the normal depth, the M_1 or M_2 profiles will then cause the depth to move *away* from the normal depth, not toward it. Thus, if the flow jumps to the 1.50 m depth from an unknown depth y, one has

$$\frac{y}{1.50} = \frac{1}{2}\left[-1 + \left(1 + 8\text{Fr}_2^2\right)^{1/2}\right]$$

with

$$\mathrm{Fr}_2^2 = \frac{q^2}{gy_2^3} = \frac{5.00^2}{9.81(1.50)^3} = 0.755$$

and the depth before the jump is $y = 1.24$ m. The correct flow sequence then has gradually varied flow beginning at the break in slope with a depth of 1.00 m and continuing downstream to a depth of 1.24 m, at which point a hydraulic jump to a depth of 1.50 m occurs.

2.25 d.

(i) The conversion of 40 acre-feet in 40 minutes to an intensity in in./h is achieved in the following way:

$$\left(\frac{40 \text{ acre-ft}}{300 \text{ acre}}\right)\left(\frac{12 \text{ in.}}{1 \text{ ft}}\right)\left(\frac{60 \text{ min/h}}{40 \text{ min}}\right) = 2.4 \text{ in./h}$$

(ii) The conversion of 1.2 inches of rain in 25 min to an intensity in in./h is direct:

$$\left(\frac{1.2 \text{ in.}}{25 \text{ min}}\right)\left(\frac{60 \text{ min}}{1 \text{ h}}\right) = 2.88 \text{ in./h}$$

The second intensity is larger.

2.26 b. In this situation, the rational method can be applied to a single land area having two uses; from Table 6.1 of *Civil Engineering PE License Review*, the neighborhood business area has a runoff coefficient $C = 0.70$, while the park has a runoff coefficient that is no larger than $C = 0.25$. The composite coefficient is

$$C = \frac{0.70(10) + 0.25(10)}{20} = 0.475$$

The rainfall intensity is 1 cm per 15 minutes, which is equivalent to 0.04 m/h, for a time interval that exceeds the time of concentration for the area. The peak discharge is therefore

$$Q_p = CiA = (0.475)\left(0.04 \frac{\text{m}}{\text{h}}\right)(10 \times 10{,}000 \text{ m}^2)\left(\frac{1 \text{ h}}{60^2 \text{ s}}\right)^2 = 0.53 \text{ m}^3/\text{s}$$

2.27 c. The storage indication method will be used in this solution. The first step is to create a plot of outflow Q versus $2S/\Delta t + Q$. The data are developed in a table:

Elev. H, ft	Storage, S_{AF}, acre-feet	Q, ft³/s	$\dfrac{2S}{\Delta t}+Q$, ft³/s
750	300	—	302
751	320	—	323
752	340	—	343
753	360	0	363
754	385	5.0	393
755	410	14.1	427
756	440	26.0	470
757	480	40.0	524

The values in the last column are computed from data in the two middle columns by the relation

$$\frac{2S}{\Delta t}+Q=\frac{2S_{AF}(43{,}560)}{24(60)^2}+Q=1.008 S_{AF}+Q \text{ ft}^3/\text{s}$$

for a 1-day, or 24-hour, routing period. Exhibit 2.27 presents a plot of these data.

Now another table will be used to carry out computations using Eq. (6.9) from *Civil Engineering PE License Review*, which is

$$I_n+I_{n+1}+\left(\frac{2}{\Delta t}S_n-Q_n\right)=\left(\frac{2S_{n+1}}{\Delta t}+Q_{n+1}\right)$$

The initial storage and outflow values are those for elevation 750, which is the initial water surface elevation in the reservoir. The values of $(2S/\Delta t+Q)_{n+1}$ are computed from Eq. (6.9); for example, for $n=1$, $(2S/\Delta t+Q)_2 = 10+25+302 = 337$ ft³/s. Since this value must exceed 363 ft³/s before outflow begins, there is no outflow at this time. When the reading of Exhibit 2.27 gives a positive outflow Q, $2Q$ is subtracted from $(2S/\Delta t+Q)_{n+1}$ to give $(2S/\Delta t-Q)_{n+1}$ in the middle column. The procedure is continued until the outflow has peaked and is receding.

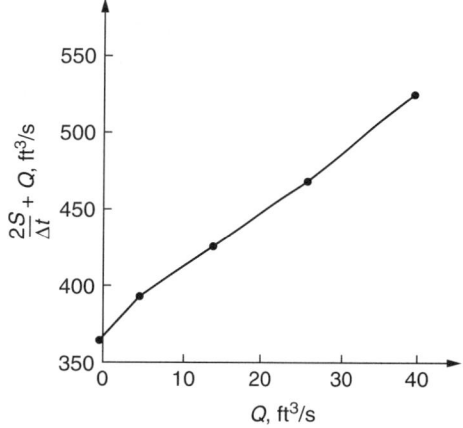

Exhibit 2.27

Time n, days	I, ft³/s	$\dfrac{2S}{\Delta t} - Q$, ft³/s	$\dfrac{2S}{\Delta t} + Q$, ft³/s	Q, ft³/s
1	10	302	—	0
2	25	337	337	0
3	35	385	397	6.0
4	30	409	450	20.5
5	25	414	464	25.0
6	20	412	459	23.5

Therefore, the peak outflow discharge is 25.0 ft³/s at the end of day 5.

2.28 b. The operating point is located at the intersection of the pump head curve and the system curve. The system curve is plotted on the pump curve in Exhibit 2.28b. From this graph, the operating point is about 2.2 cfs.

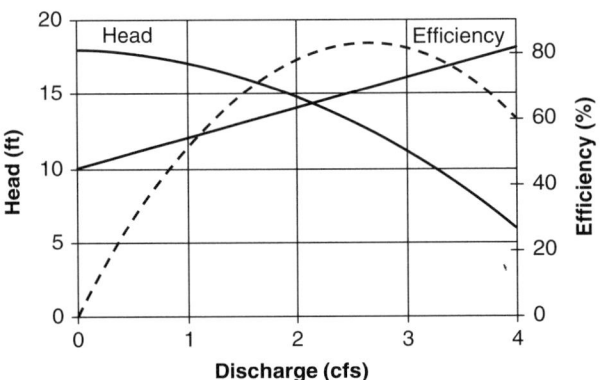

Exhibit 2.28b

2.29 c. The full pipe velocity is given by Q/A, or $(1 \text{ cfs})/(\pi \times (0.5 \text{ ft})^2)$, or 1.3 fps. By definition, or inspection of a hydraulic elements chart, the half-full velocity equals the full-pipe velocity, and thus the half-full flow is half of the full-pipe flow.

2.30 b. The volume of runoff is the area under the hydrograph. This total area has been estimated in Exhibit 2.30b, assuming a base flow of 20 cfs, as the sum of Triangle A and Triangle B.

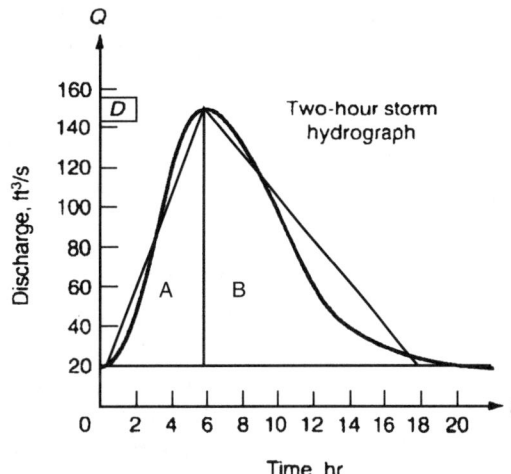

Exhibit 2.30b

$$A_{\text{total}} = A_A + A_B$$
$$= 0.5 \times (6 \text{ hrs}) \times (150 \text{ cfs} - 20 \text{ cfs}) \times (3600\text{s/h})$$
$$+ 0.5 \times (12 \text{ h}) \times (150 \text{ cfs} - 20 \text{ cfs}) \times (3600\text{s/h})$$
$$= 4.2 \times 10^6 \text{ ft}^3$$

2.31 a. The head losses around any given loop must sum to zero. Traveling around loop 2-3-4-1 shows that the head loss in pipe 3-4 must equal 1 ft. This answer is obtained as follows:

$$h_{L,2-3} + h_{L,3-4} - h_{L,4-1} + h_{L,1-2} = 0$$

The signs on each term are assigned using this convention: positive in the direction of flow, negative against the direction of flow.

Inserting the known values, we arrive at

$$3 \text{ ft} + h_{L,3-4} - 5 \text{ ft} + 1 \text{ ft} = 0, \quad \text{or} \quad h_{L3-4} = 1 \text{ ft}$$

2.32 d. The flow in a pipe is related to the pipe's diameter, length, and head loss. Even though the head losses in the two pipes are equal and the diameters are the same, we do not know anything about the pipes' lengths. Without knowing the pipe length, we cannot make any estimation of the flow rate in an individual pipe.

2.33 b. THMs (trihalomethanes) are an example of a disinfection by-product.

2.34 b. The volume of sludge sent to disposal equals the mass of residuals exiting the filter press multiplied by the specific weight of the dewatered residuals. Thus, this is a two-part problem: first find the specific weight of the dewatered residuals, and then find the mass of residuals exiting the filter press.

The specific weight of the dewatered residuals is a weighted average of the specific weight of the water and of the solids. So, for a 100-gram sample of dewatered sludge (consisting of 20 grams of solids and 80 grams of water), recalling that the specific weight of water is 62.4 lb/ft³, the specific weight of the sludge is given as

$$\gamma_{sludge} = \frac{(20\,g)\left(2.0 \cdot 62.4\,\frac{lb}{ft^3}\right) + (80\,g)\left(62.4\,\frac{lb}{ft^3}\right)}{100\,g} = 75\,\frac{lb}{ft^3}$$

To determine the mass of residuals exiting the filter press, consider that every 100,000 pounds of sludge sent to the dewatering device contains 2000 pounds of solids. Moreover, if we assume 100 percent capture of the solids, the sludge exiting the dewatering device will also contain 2000 pounds of solids. Given the definition of total solids, the mass of sludge exiting the filter press can be determined:

$$TS = \frac{mass\ of\ solids}{mass\ of\ sludge} \quad or \quad 20\% = \frac{2000\ lb}{x}$$

so x (the mass of dewatered sludge) equals 10,000 lb. The daily volume of sludge sent to disposal then equals (10,000 lb)/(75 lb/ft³) = 133 ft³, or 995 gallons.

2.35 b. Given the definition of detention time, volume divided by flow rate, and the known flow rate, we simply need to find the detention time. This time is given by a mass balance approach to a batch reactor. In other words,

$$C_t = C_0 e^{-kt}$$

where C_t is the concentration in the batch reactor at any time t, C_0 is the initial concentration, k is the decay constant, and t is time.

The required 99.9 percent removal corresponds to $C_t/C_0 = 1/1000$. Solving the mass balance equation for time yields

$$t = \frac{\ln(1/1000)}{-k}, \quad or \quad t = 35\ minutes$$

From the detention time equation, the volume required is then 2.4×10^4 gallons.

2.36 d. Land development decreases interception by decreasing the area of foliage that can intercept rainfall before it hits the ground.

Infiltration is decreased by land development due to the exchange of pervious surface for impervious surfaces.

Transpiration is decreased by land development due to the elimination of vegetation, which is responsible for transpiration.

2.37 d. The saturated aquifer thicknesses at the observation wells are $h_1 = 50 - 18 = 42$ m and $h_2 = 50 - 7 = 43$ m. The hydraulic conductivity is $K = (0.025 \text{ cm/s})(1 \text{ m}/100 \text{ cm}) = 0.00025$ m/s. The insertion of these data in Eq. (6.18) of *Civil Engineering PE License Review* results in a discharge

$$Q = \frac{\pi K \left[h_2^2 - h_1^2 \right]}{\ln(r_2/r_1)} = \frac{\pi (0.00025)[(43)^2 - (42)^2]}{\ln(45/20)} = 0.0823 \text{ m}^3/\text{s}$$

Hence, the discharge is about 82 liters/s.

2.38 c. Given that the overflow rate = (flow)/(area), the area of the tank is

$$\text{area} = \text{flow/overflow rate}$$
$$= (0.2 \text{ m}^3/\text{s})/(20 \text{ m/day})$$
$$= 864 \text{ m}^2$$

The volume of the tank is the tank area multiplied by the tank depth.

$$\text{volume} = \text{area} \times \text{depth} = 864 \text{ m}^2 \times 2.5 \text{ m} = 2160 \text{ m}^3$$

Finally, the detention time θ is

$$\theta = \text{volume/flow} = (2160 \text{ m}^3)/(0.2 \text{ m}^3/\text{s}) = 180 \text{ minutes}$$

2.39 a. The answer is found by calculating the Hazen-Williams C value from the given information. The pipe material can then be deduced from the C value.

The Hazen-Williams equation is

$$h_L = \frac{4.73 L}{C^{1.852} D^{4.87}} \times Q^{1.852} \quad \text{for } L \text{ in ft, } D \text{ in ft, and } Q \text{ in cfs}$$

Given

$$h_L = [(60 \text{ psi} - 50 \text{ psi}) \times (144 \text{ in.}^2/\text{ft}^2)]/(62.4 \text{ lbf/ft}^3) = 23 \text{ ft}$$
$$L = 800 \text{ feet}$$
$$D = 0.5 \text{ feet}$$
$$Q = 1.57 \text{ cfs}$$

Solving the Hazen-Williams equation for C yields a C value of around 152, which suggests that the pipe is PVC.

2.40 a. The intensity corresponds to the time of concentration (t_c) to Inlet Y. By inspection, t_c for Inlet Y is the time to travel within Catchment X plus the time of travel (t_t) in the pipe between Inlet X and Inlet Y.

The time of travel is the pipe length divided by the pipe velocity. Pipe velocity can be estimated by Manning's Equation, assuming full pipe flow.

$$V = 1.486 n^{-1} (R_h)^{2/3} (S_o)^{1/2}$$

where
> R_h for an 18-inch circular cross section is $D/4$, or (18 inches)/4, or 0.375 ft
> $n = 0.013$ for concrete pipe
> $S_o = 0.02$

This yields a V of 8.41 ft/s and a t_t of (1000 ft)/(8.4 ft/s) = 120 min.

Given this t_c (120 min + 10 min = 130 min), the intensity from the provided intensity-duration curve is estimated to be around 1 in./h.

SOLUTIONS TO CHAPTER 3 PROBLEMS

3.1 b.

$$\text{Moist unit weight,} \gamma = \frac{W}{V} = \frac{18}{0.15} = 120 \text{ lb/ft}^3$$

$$\text{Dry unit weight,} \gamma_d = \frac{\gamma}{1+\frac{w}{100}} = \frac{120}{1+\frac{12}{100}} = 107.1 \text{ lb/ft}^3$$

$$\text{Void ratio,} e = \frac{G_s \gamma_w}{\gamma_d} - 1 = \frac{(2.69)(62.4)}{107.1} - 1 = 0.567$$

$$\text{Degree of saturation,} S = \frac{wG_s}{e} = \frac{(0.12)(2.69)}{0.567} \times 100 = 56.9\%$$

3.2 b.

$$\gamma_{\text{in situ}} = 102 \text{ lb/ft}^3$$

$$\gamma_{d \text{ in situ}} = \frac{102}{1+\frac{16}{100}} = 87.93 \text{ lb/ft}^3$$

$$\text{Volume to be excavated} = (8000)\left(\frac{103.5}{87.93}\right) = 9417 \text{ yd}^3$$

3.3 c.

Given: $\rho_{\text{sat(clay)}} = 1925$ kg/m³. Thus,

$$\gamma_{\text{sat(clay)}} = \frac{(1925)(9.81)}{1000} = 18.88 \text{ kN/m}^3$$

Due to excavation, there will be unloading of the overburden pressure. Let the depth of the cut be H, at which point the bottom will heave. Let us consider the stability of point A at that time.

$$\sigma_A = (10 - H)\gamma_{sat(clay)}$$
$$u_A = 6\gamma_w$$

For heave to occur, σ'_a should be 0. So

$$\sigma_A - u_A = (10 - H)\gamma_{sat(clay)} - 6\gamma_w$$

or

$$(10 - H)18.88 - (6)9.81 = 0$$

$$H = \frac{(10)18.88 - (6)9.81}{18.88} = 6.88 \text{ m}$$

3.4 b.

$$m_v = \frac{a_v}{a + e_{av}} = \frac{\frac{\Delta e}{\Delta p}}{1 + e_{av}}$$

$$e_{av} = \frac{0.72 + 0.6}{2} = 0.66$$

$$m_v = \left(\frac{0.72 - 0.6}{700 - 350}\right)\frac{1}{1 + 0.66} = 2.07 \times 10^{-4} \text{ m}^2/\text{kN}$$

3.5 c. Refer to Fig. 9.9 in *Civil Engineering PE License Review*.

$$\frac{\text{Length}}{\text{Width}} = \frac{L}{B} = 1$$

$$\frac{z}{B} = \frac{4}{2} = 2$$

For $\frac{L}{B} = 1$ and $\frac{z}{B} = 2$, the magnitude of $\frac{\Delta p}{q} \approx 0.1$. So

$$\Delta p = (0.1)(200) = 20 \text{ kN/m}^2$$

3.6 c. Consolidation settlement

$$S = \frac{C_c H}{1 + e_o} \log\left(\frac{p_o + \Delta p_{av}}{p_o}\right)$$

$$= \frac{(0.3)(3)}{1 + 0.9} \log\left(\frac{90.59 + 27}{90.59}\right) = 0.0537 \text{ m} = 53.7 \text{ mm}$$

3.7 c. Effective major principal stress = $\sigma'_1 = 10 + 8 - 6 = 12$ lb/in.^{2}

Effective minor principal stress = $\sigma'_3 = 10 - 6 = 4$ lb/in.^{2}

$$\sigma_1' = \sigma_3' \tan^2\left(45° + \frac{\phi}{2}\right)$$

$$12 = 4\tan^2\left(45° + \frac{\phi}{2}\right); \phi = 30°$$

3.8 b. Refer to Fig. 9.14 in *Civil Engineering PE License Review*. The depth of tensile crack, z, is where $\sigma_A = 0$. So, from Eq. (9.40),

$$\sigma_o = 0 = K_a \sigma_v - 2c_u \sqrt{K_a}$$

$$K_a = \tan^2\left(45° - \frac{\phi}{2}\right) = \tan^2\left(45° - \frac{0°}{2}\right) = 1$$

So,

$$\gamma z - 2c_u = 0$$

$$z = \frac{2c_u}{\gamma} = \frac{(2)(20)}{19} = 2.1 \text{ m}$$

$$\sigma_{a(z=H)} = \gamma H - 2c_u \text{ (with } K_a = 1\text{)} = (19)(6) - (2)(20) = 74 \text{ kN/m}^2$$

$$P_a = \frac{1}{2}(74)(6 - 2.1) = 144.3 \text{ kN/m}$$

3.9 c.

$$\text{Area ratio, } A_r = \frac{D_o^2 - D_i^2}{D_i^2} \times 100 = \frac{(50.8)^2 - (34.93)^2}{(34.93)^2} \times 100 = 111.5\%$$

3.10 c.

$$K_a = \tan^2\left(45° - \frac{\phi}{2}\right) = \tan^2\left(45° - \frac{30}{2}\right) = \frac{1}{3}$$

Pressure distribution:

At $z = 0$: $\sigma_a' = \sigma_v' K_a = (0)\left(\frac{1}{3}\right) = 0$

$u = 0$

At $z = 2$ m: $\sigma_a' = (15.7)(2)\left(\frac{1}{3}\right) = 10.47 \text{ kN/m}^2$

$u = 0$

At $z = 4$ m: $\sigma_a' = [(15.7)(2) + (18.2 - 9.81)(2)]\left(\frac{1}{3}\right) = 16.06 \text{ kN/m}^2$

$u = (9.81)(2) = 19.62 \text{ kN/m}^2$

Refer to Exhibit 3.10b.

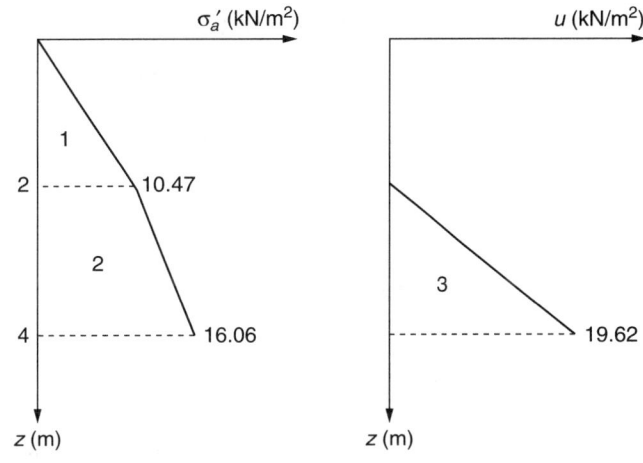

Exhibit 3.10b

$$\text{Area 1} = \left(\frac{1}{2}\right)(2)(10.47) = 10.47$$

$$\text{Area 2} = \left(\frac{10.47 + 16.06}{2}\right)(2) = 26.53$$

$$\text{Area 3} = \left(\frac{1}{2}\right)(2)(19.62) = 19.62$$

$$\text{Area 1 + Area 2 + Area 3} = 56.62 \text{ kN/m}$$

3.11 c.

$$P_a = \frac{1}{2}\gamma H'^2 K_a$$

$$K_a = \tan^2\left(45° - \frac{\phi}{2}\right) = \tan^2\left(45° - \frac{30}{2}\right) = \frac{1}{3}$$

$$P_a = \left(\frac{1}{2}\right)(105)(21.5)^2\left(\frac{1}{3}\right) = 8089.3 \text{ lb/ft} \approx 8090 \text{ lb/ft}$$

The following table can now be prepared. Refer to Exhibit 3.11.

Section No.	Weight, lb/ft	Moment Arm From c, ft	Moment About c, lb-ft/ft
1	20 × 7.5 × 105 = 15,750	10.25	161,437.5
2	20 × 1.5 × 150 = 4500	5.75	25,875
3	14 × 1.5 × 150 = 3150	7	22,050
	ΣV = 23,400 lb/ft		M_R = 209,362 lb-ft/ft

$$M_O = (P_a)(7.17) = (8090)(7.17) = 58,005.3 \text{ lb-ft/ft}$$

$$\text{FS}_{(\text{overturning})} = \frac{209,362.5}{58,005.3} = 3.61$$

3.12 c.

Point load bearing capacity $Q_p = 9c_u A_p = (9)(1000)(1.5)^2 = 20{,}250$ lb

Side load $Q_s = \alpha c_u pL = (0.75)(1000)(4 \times 1.5)(100) = 450{,}000$ lb

Total load carrying capacity $= Q_p + Q_s = 470{,}250$ lb

3.13 c.

$$\text{RQD} = \frac{1.7 \text{ ft}}{3 \text{ ft}} \times 100 = 56.7\%$$

3.14 b.

$$\gamma_{sat} = \frac{(G_s + e)\gamma_w}{1+e}$$

where γ_w = unit weight of water

$$\gamma_{sat} = \frac{(2.68 + 0.71)(62.4)}{1 + 0.71} = 123.7 \text{ lb/ft}^3$$

3.15 a.

$$\gamma_d = \frac{G_s \gamma_w}{1+e}$$

$$105 = \frac{(2.68)(62.4)}{1+e}$$

$$e = 0.59$$

For quicksand conditions:

$$i_{cr} = \frac{\gamma'}{\gamma_w} = \frac{G_s - 1}{1+e} = \frac{2.68 - 1}{1 + 0.59} = 1.06$$

3.16 b.

$$q_{II} = \frac{k(H_1 - H_2)}{N_d}$$

where N_d is the number of potential drops

$$q_{II} = \frac{(0.08 \times 10^{-2} \text{ m/s})(3.96 - 0.53)}{6} = 4.57 \times 10^{-4} \text{ m}^3/\text{s/m}$$

3.17 c.

$$q_{total} = k(H - H_2)\frac{N_f}{N_d}$$

where N_f is the number of flow channels

$$Q_{total} = (0.08 \times 10^{-2} \text{ m/s})(3.96 - 0.53)\frac{4}{6} = 1.83 \times 10^{-3} \text{ m/s/m}$$

3.18 b. The head loss between *a* and *b* is

$$\Delta h = \frac{H_1 - H_2}{N_d} = \frac{3.96 - 0.53}{6} = 0.572 \text{ m}$$

The length of *ab* ≈ 1.98 m.

$$i = \frac{\Delta h}{\text{length}} = \frac{0.572}{1.98} = 0.289$$

3.19 a.

$$\sigma_1 = \sigma_3 + \Delta\sigma_d = 70 + 120 = 190 \text{ kN/m}^2$$
$$\sigma_1 = \sigma_1'; \; \sigma_3 = \sigma_3'$$
$$\frac{\sigma_1'}{\sigma_3'} = \tan^2\left(45 + \frac{\phi}{2}\right)$$
$$\frac{190}{70} = \tan^2\left(45 + \frac{\phi}{2}\right)$$
$$\phi = 27.5°$$

3.20 c. The area ratio is given by

$$A_r \text{ (\%)} = \frac{D_o^2 - D_i^2}{D_i^2} = \frac{(2)^2 - (1.875)^2}{(1.875)^2} \times 100 = 13.8\%$$

3.21 b. The Rankine active pressure coefficient can be calculated as

$$K_a = \tan^2\left(45 - \frac{\phi}{2}\right) = \tan^2\left(45 - \frac{36}{2}\right) = 0.26$$

Thus, the intensity of active pressure is

$$\sigma_a = \gamma z K_a = (110)(10)(0.26) = 286 \text{ lb/ft}^2$$

3.22 a. Find the active force:

$$P_a = \tfrac{1}{2}\gamma H^2 K_a = (\tfrac{1}{2})(110)(10)^2(0.26) = 1430 \text{ lb/ft}$$

3.23 c. The Rankine passive pressure coefficient can be calculated as

$$K_p = \tan^2\left(45 + \frac{\phi}{2}\right) = \tan^2\left(45 + \frac{36}{2}\right) = 3.85$$

The passive force is

$$P_p = \tfrac{1}{2}\gamma H^2 K_p = (\tfrac{1}{2})(110)(10)^2(3.85) = 21{,}175 \text{ lb/ft} \approx 21.2 \text{ kip/ft}$$

3.24 c.

$$S = \frac{C_c H}{1+e_o} \log\frac{p_o + \Delta p}{p_o} = \frac{(0.27)(10 \times 12)}{1+1.1}\log\left(\frac{1600+1000}{1600}\right) = 3.25 \text{ in.}$$

3.25 c.

$$C_c = \frac{e_1 - e_2}{\log\dfrac{p_2}{p_1}} = \frac{1.1 - 0.98}{\log\dfrac{4000}{2000}} = 0.399$$

3.26 b.

$$\sigma_a = 0.65\gamma H K_a = 0.65\gamma H \tan^2\left(45 - \frac{\phi}{2}\right)$$

$$= (0.65)(16)(9)\tan^2\left(45 - \frac{30}{2}\right) = 31.2 \text{ kN/m}^2$$

Refer to Exhibit 3.26b.

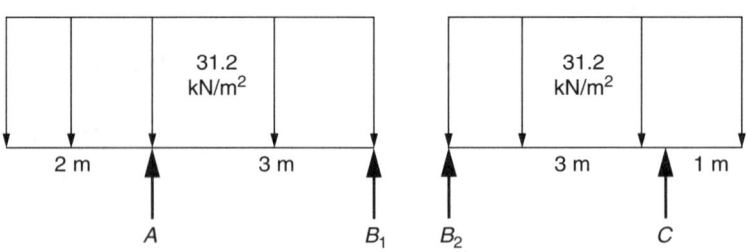

Exhibit 3.26b

$$\sum M_{B_1} = (A)(3) - (31.2)(5)\left(\frac{5}{2}\right) = 0$$

$$A = 130 \text{ kN}$$

Strut load at A = (130)(spacing of struts) = (130)(3) = 390 kN.

3.27 b. The equivalent eccentricity of the moment acting on the pile group is

$$e = M/W$$
$$= 50/50$$
$$= 1 \text{ foot}$$

The line of piles is symmetrical with an inertia of

$$\Sigma x^2 = 2(10^2)$$
$$= 200$$

The maximum axial force occurs in the right-hand pile and, neglecting its self-weight, is given by

$$P = W/n_L + Wex/\Sigma x^2$$
$$= 50/3 + 50 \times 1 \times 10/200$$
$$= 19.2 \text{ kips}$$

3.28 c. The relevant properties of the column are

$$b = \text{effective column width}$$
$$= 15.63 \text{ inches}$$
$$h = \text{effective column height}$$
$$= 40 \text{ feet}$$
$$A_{st} = \text{reinforcement area}$$
$$= 2.40 \text{ in}^2$$
$$A_n = \text{effective column area}$$
$$= b^2$$
$$= (15.63 \text{ in.})^2$$
$$= 244 \text{ in.}^2$$

The radius of gyration of the column is given by BCRMS Commentary Section 1.9.3 as

$$r = (I_n/A_n)^{0.5}$$
$$= 0.289b$$
$$= 4.52 \text{ in.}$$

The slenderness ratio of the column is

$$h/r = 40 \times 12/4.52$$
$$= 106.20$$
$$> 99 \ldots \text{BCRMS Equation (2-18) is applicable}$$

The allowable axial load is given by

$$P_a = (0.25 f'_m A_n + 0.65 A_{st} F_s)(70r/h)^2$$
$$= (0.25 \times 1.5 \times 244 + 0.65 \times 2.40 \times 24)(70/106.20)^2$$
$$= 56 \text{ kips}$$

3.29 a. The net factored loading on the footing is

$$q_u = P_u/A_f$$
$$= 250/(8 \times 8)$$
$$= 3.91 \text{ ksf}$$

The critical section for punching shear in a footing, as defined by ACI Sections 15.5.2 and 11.12.1.2, is located on the perimeter of a square with sides of length

$$b = c + d$$
$$= 12 + 15$$
$$= 27 \text{ inches}$$

The length of the critical perimeter is

$$b_o = 4b$$
$$= 4 \times 27$$
$$= 108 \text{ inches}$$

The ratio of the long side to the short side of the column is

$$\beta = c/c$$
$$= 12/12$$
$$= 1.0$$
$$< 2$$

Hence, the punching shear capacity of the footing is given by ACI Equation (11–35) as

$$\phi V_c = 4\phi b_o d(f'_c)^{0.5}$$
$$= 4 \times 0.75 \times 108 \times 15(3000)^{0.5}/1000$$
$$= 266 \text{ kips}$$

3.30 b. As specified by BCRMS Section 2.1.6.5, lateral tie spacing shall not exceed the lesser of

$$s = 48 \text{ lateral tie diameters}$$
$$= 48 \times 0.375$$
$$= 18 \text{ inches}$$
$$s = \text{the least cross-sectional dimension of the column}$$
$$= 16 \text{ inches}$$
$$s = 16 \text{ longitudinal bar diameters}$$
$$= 16 \times 0.875$$
$$= 14 \text{ inches } ... \text{ governs}$$

3.31 c. The net factored loading on the footing is

$$q_u = P_u/A_f$$
$$= 300/(8 \times 8)$$
$$= 4.69 \text{ ksf}$$

The critical section for flexure in a footing, as defined by ACI Section 15.4.2, is located along the side of the column. The factored applied moment at this section is

$$M_u = q_u l(l/2 - c/2)^2/2$$
$$= 4.69 \times 8(8/2 - 1/2)^2/2$$
$$= 230 \text{ kip-feet}$$

The required reinforcement ratio, assuming a tension-controlled section, is derived from ACI Section 10.2, and is

$$\rho = 0.85 f_c' [1 - (1 - 2M_u/0.765bd^2 f_c')^{0.5}]/f_y$$
$$= 0.00213$$

The required reinforcement area, assuming a tension-controlled section, is

$$A_s = \rho l d$$
$$= 0.00213 \times 96 \times 16$$
$$= 3.27 \text{ in.}^2$$

The maximum allowable reinforcement ratio for a tension-controlled section is given by ACI Section 10.3.4 as

$$\rho_t = 0.319 \beta_1 f_c'/f_y$$
$$= 0.319 \times 0.85 \times 3/60$$
$$= 0.0136$$
$$> \rho \ldots \text{ satisfactory}$$

The minimum reinforcement area in a footing slab for Grade 60 bars is given by ACI Section 7.12.2.1 as

$$A_{s(min)} = 0.18\% \text{ of the gross area}$$
$$= 0.0018 \times l \times h$$
$$= 0.0018 \times 96 \times 19$$
$$= 3.28 \text{ square inches}$$
$$> A_s \ldots \text{ governs}$$

3.32 b. The nominal $3/4$-inch-diameter bolt design value in a 4 × Douglas Fir-Larch sawn lumber member, for single shear perpendicular to grain, into concrete is tabulated in NDS Table 11E as

$$Z_\perp = 1070 \text{ lb}$$

The maximum vertical load on the ledger is

$$w = 1070/4$$
$$= 268 \text{ pounds/foot}$$

3.33 b. The maximum factored bending moment per foot length of stem is

$$M_u = 1.6 \times 30 \times 18.5^3/6$$
$$= 50{,}653 \text{ pound-feet}$$

The required reinforcement ratio, assuming a tension-controlled section, is derived from ACI Section 10.2 and is

$$\rho = 0.85 f'_c [1 - (1 - 2M_u/0.765bd^2 f'_c)^{0.5}]/f_y$$
$$= 0.0044$$

The required reinforcement area, assuming a tension-controlled section, is

$$A_s = \rho l d$$
$$= 0.0044 \times 12 \times 15$$
$$= 0.79 \text{ in.}^2$$

The maximum allowable reinforcement ratio for a tension-controlled section is given by ACI Section 10.3.4 as

$$\rho_t = 0.319 \beta_1 f'_c / f_y$$
$$= 0.319 \times 0.85 \times 3/60$$
$$= 0.0136$$
$$> \rho \ldots \text{ satisfactory}$$

3.34 d. The applied horizontal force on the retaining wall is

$$H = 30 \times 20^2/2$$
$$= 6000 \text{ pounds}$$

The frictional resistance available is

$$F = \mu \Sigma W$$
$$= 0.4 \times 20{,}000$$
$$= 8000 \text{ pounds}$$

To provide a factor of safety of 1.5 against sliding requires a passive resistance of

$$H_P = 1.5H - F$$
$$= 1.5 \times 6000 - 8000$$
$$= 1000 \text{ pounds}$$

To provide this passive force, the total required depth of the shear key plus footing is

$$H_K = (2H_P/p_P)^{0.5}$$
$$= (2 \times 1000/300)^{0.5}$$
$$= 2.6 \text{ feet}$$

The depth of the shear key is

$$D = H_K - 1.5$$
$$= 1.1 \text{ feet}$$

3.35 c. Volumes between the stations for which cross section areas are given may be calculated by

Station 50 + 00 to 50 + 40: $V = (120.25 + 60.75)(40/54) = 134.1$ yd^3
Station 50 + 40 to 51 + 20: $V = (60.75 + 15.87)(80/54) = 113.5$ yd^3

The total excavation is $134.1 + 113.5 = 247.6$ yd^3.

3.36 d.

The area is given by

$$A = \left| \sum_{i=1}^{n} [X_i(Y_{i+1} - Y_{i-1})]/2 \right|$$

$$= \big|[(-6)(2-0) + (-14)(2.4-0) + (-8)(2.5-2.0) + 0(2.3-2.4)$$

$$+ 10(2.2-2.5) + 14.8(0-2.3) + 6(0-2.2)]/2\big|$$

$$= \big|[-12 - 33.6 - 4.0 + 0 - 3.0 - 34.04 - 13.2]/2\big| = 49.92 \text{ m}^2$$

3.37 a. The hydraulic conductivity of an unconfined aquifer can be calculated from

$$K = \frac{Q \ln\left(\frac{r_2}{r_1}\right)}{\pi\left(h_2^2 - h_1^2\right)}$$

Values for Q, r_1, and r_2 were given in the problem statement along with the drawdown in each well. Water depth in each well is determined by subtracting the drawdown from the aquifer thickness. Aquifer thickness can be found by taking the difference between the depth to the confining layer and the depth to the groundwater table, which is 40 ft – 12 ft, or 28 ft thick. Water depth in each well can now be determined as follows:

$$h_1 = 28 - 4.6 = 23.4 \text{ ft} \quad \text{and} \quad h_2 = 28 - 0.3 = 27.7 \text{ ft}$$

Therefore, the hydraulic conductivity of the aquifer can be determined as follows:

$$K = \frac{(5 \text{ gpm}) \ln\left(\frac{40}{30}\right)}{\pi[(27.7 \text{ ft})^2 - (23.4 \text{ ft})^2]} = 0.002084 \, \frac{\text{gpm}}{\text{ft}^2} \times \frac{1 \text{ ft}^3}{7.48 \text{ gal}}$$

$$= 0.000279 \, \frac{\text{ft}}{\text{min}}$$

$$\left(0.000279 \, \frac{\text{ft}}{\text{min}}\right)\left(60 \, \frac{\text{min}}{\text{h}}\right)\left(24 \, \frac{\text{h}}{\text{day}}\right)\left(365 \, \frac{\text{day}}{\text{yr}}\right) = 146.4 \, \frac{\text{ft}}{\text{yr}}$$

3.38 b. First, determine the volume of affected water and soil as follows:

$$(100 \text{ ft})(600 \text{ ft})(28 \text{ ft}) = 1.68 \times 10^6 \text{ ft}^3 \times 0.32 = 537{,}600 \text{ ft}^3 \text{ H}_2\text{O}$$
$$\text{and} \quad 1.68 \times 10^6 \text{ ft}^3 \times 0.68 = 1{,}142{,}400 \text{ ft}^3 \text{ soil}$$

Determine the total mass of contaminant in the aqueous phase as follows:

$$(537{,}600 \text{ ft}^3)\left(28.3 \frac{\text{L}}{\text{ft}^3}\right)\left(2.3 \frac{\text{mg contaminant}}{\text{L}}\right)\left(10^{-6} \frac{\text{kg}}{\text{mg}}\right)$$
$$= 35.0 \text{ kg contaminant}$$

To obtain the mass of contaminant associated with the soil phase, first determine the soil phase concentration using the soil-water partition coefficient as follows:

$$K_{SW} = \frac{C_{soil}}{C_{aq}} \quad \Rightarrow \quad C_{soil} = K_{SW} \times C_{aq} = (0.4)(2.3) = 0.92 \frac{\text{mg}}{\text{kg}}$$

Then determine the total mass of contaminant in the soil phase as follows:

$$(1{,}142{,}400 \text{ ft}^3)\left[(2.7)(62.4) \frac{\text{lb soil}}{\text{ft}^3 \text{ soil}}\right]\left(\frac{1 \text{ kg}}{2.2 \text{ lb}}\right)\left(0.92 \frac{\text{mg contaminant}}{\text{kg soil}}\right)$$
$$\times \left(10^{-6} \frac{\text{kg}}{\text{mg}}\right) = 80.5 \text{ kg contaminant}$$

Therefore, the total mass of contaminant is

$$35 + 80.5 = 115.5 \text{ kg} \times \left(2.2 \frac{\text{lb}}{\text{kg}}\right) = 254 \text{ lb contaminant}$$

Now the volume can be calculated from the specific weight of the contaminant, which is calculated from the specific gravity given as follows:

$$\gamma_{contaminant} = (0.87)\left(62.4 \frac{\text{lb}}{\text{ft}^3}\right) = 54.3 \frac{\text{lb}}{\text{ft}^3}$$

$$\frac{254 \text{ lb}}{54.3 \frac{\text{lb}}{\text{ft}^3}} = 4.68 \text{ ft}^3 \times \left(7.48 \frac{\text{gal}}{\text{ft}^3}\right) = 35.0 \text{ gallons}$$

3.39 d. The fraction of organic matter in the soil can be calculated from the following equation:

$$K_{sw} = K_{oc} f_{oc}$$

Since K_{sw} was given in the problem statement, this requires only calculating K_{oc}. K_{oc} may be estimated from K_{ow} using the following relationship:

$$\log K_{oc} = \log K_{ow} - 0.21$$

K_{ow} may be calculated from the data in the problem statement as follows:

$$K_{ow} = \frac{C_o}{C_w} = \frac{123}{4.6} = 26.74$$

Now use this value to estimate K_{oc} as follows:

$$\log K_{oc} = \log(26.74) - 0.21 = 1.427 - 0.21 = 1.217$$
$$K_{oc} = 10^{1.217} = 16.48$$

f_{oc} may now be calculated as follows:

$$f_{oc} = \frac{K_{sw}}{K_{oc}} = \frac{0.4}{16.48} = 0.0243 \quad \text{or} \quad f_{oc} = 2.43\%$$

3.40 d. The hydraulic conductivity calculated in Problem 3.37 gives the water velocity. However, the contaminant is slowed by its affinity for sorption onto the organic fraction of the soil. This slowing is referred to as "retardation" and may be estimated by the following equation:

$$R = 1 + \frac{\rho}{\eta} K_{sw}$$

K_{sw} and soil porosity are already known, leaving the bulk density to be determined to calculate the retardation factor. The bulk density may be determined knowing the fraction of water and soil and the density (or specific gravity) of both as follows:

$$\rho = (0.32)\left(1 \, \frac{g}{cm^3}\right) + (0.68)(2.7)\left(1 \, \frac{g}{cm^3}\right) = 2.156 \, \frac{g}{cm^3}$$

Now the retardation factor can be calculated from

$$R = 1 + \left(\frac{2.156}{0.32}\right)(0.4) = 3.7$$

Because the retardation factor expresses the relative velocity of the water to the contaminant, the velocity of the contaminant through the aquifer may be determined as follows:

$$R = \frac{V_{water}}{V_{contaminant}} = 3.7 \quad \text{or} \quad V_{contaminant} = \frac{V_{water}}{R} = \frac{146.4 \, \frac{\text{ft}}{\text{yr}}}{3.7} = 39.6 \, \frac{\text{ft}}{\text{yr}}$$

Therefore, the time for the contaminant to reach the well may be determined as follows:

$$\frac{(300 \text{ yd})\left(3 \, \frac{\text{ft}}{\text{yd}}\right)}{39.6 \, \frac{\text{ft}}{\text{yr}}} = 22.7 \text{ years}$$

SOLUTIONS TO CHAPTER 4 PROBLEMS

4.1 c. From AASHTOSD Section 3.3, the importance classification for a bridge located on a strategic route is

$$IC = I$$

From AASHTOSD Section 3.4, for an importance classification of I and an acceleration coefficient exceeding 0.29, the seismic performance category is

$$SPC = D$$

From AASHTOSD Section 4.2, for a regular bridge with three spans and a seismic performance category of D, the required analysis procedure is Procedure 1 or 2.

4.2 b.

ASD Option

In the absence of wind or seismic loads, the load factor specified by AISC Section N1 is

$$n = 1.7$$

At collapse, plastic hinges form at A, B, and C. Taking moments about B for segment AB gives

$$\begin{aligned} 2M_P &= 1.7[R_A L/2 - w(L/2)(L/4)] \\ &= 1.7(150 \times 20/2 - 10 \times 20^2/8) \\ &= 1700 \\ M_P &= 850 \text{ kip-ft} \end{aligned}$$

From AISC, a W30 × 108 has a plastic moment of

$$\begin{aligned} M_p &= 863 \text{ kip-ft} \\ &> 850 \text{ kip-ft} \ldots \text{satisfactory} \end{aligned}$$

LRFD Option

The factored loads are given by AISC, 13th ed., Part 2, and IBC Eq. (16-2) as

$$W_u = 1.6 \times 100$$
$$= 160 \text{ kips}$$
$$w_u = 1.2 \times 10$$
$$= 12 \text{ klf}$$

The bending moment at support A governs and is given by

$$M_u = W_u \times 20/8 + w_u \times 20^2/12$$
$$= 160 \times 20/8 + 12 \times 20^2/12$$
$$= 800 \text{ kip-ft}$$

The required plastic section modulus is

$$Z = M_u/\phi F_y$$
$$= 800 \times 12/(0.9 \times 50)$$
$$= 213.3 \text{ in.}^3$$

From AISC, 13th ed., Table 3-2, a W30 × 90 has a plastic section modulus of

$$Z = 283 \text{ in.}^3$$
$$> 213.3 \text{ in.}^3 \ldots \text{satisfactory}$$

4.3 d. The shear force along the east wall is

$$V = pL/2$$
$$= 280 \times 60/2$$
$$= 8400 \text{ lb}$$

The unit shear along the east wall is

$$q = V/B$$
$$= 8400/20$$
$$= 420 \text{ lb/ft}$$

The maximum service-level force in the drag strut CD is

$$F = qL_{CD}$$
$$= 420 \times 10$$
$$= 4200 \text{ lb}$$

In accordance with ASCE Section 9.5.2.6.3.1, a light-framed structure may be designed for this force.

From NDS Table 2.3.2, the load duration factor for seismic loading is

$$C_D = 1.6$$

From NDS Table 11N, the nail design value for a 16d common nail in single shear, for Douglas Fir–Larch, is

$$Z = 141 \text{ lb}$$

The number of nails required is

$$\begin{aligned} n &= F/ZC_D \\ &= 4200/(141 \times 1.6) \\ &= 19 \text{ nails} \end{aligned}$$

4.4 a. The beam self-weight is

$$\begin{aligned} w_s &= 24 \times 12 \times 150/144 \\ &= 0.3 \text{ klf} \end{aligned}$$

From ACI Eq. (9-2), the factored load is

$$\begin{aligned} w_u &= 1.2w_D + 1.6w_L \\ &= 1.2(1.0 + 0.3) + 1.6 \times 1 \\ &= 3.16 \text{ klf} \end{aligned}$$

From ACI Section 8.3.3, the factored positive moment in the end span is

$$\begin{aligned} M_u &= w_u \ell_n^2/14 \\ &= 3.16 \times 35^2/14 \\ &= 276.5 \text{ kip-ft} \end{aligned}$$

4.5 c. For a site with an undetermined soil profile, the site classification specified by ASCE Section 11.4.2 is D.

From ASCE Table 11.4-1, the site coefficient is

$$F_a = 1.2$$

From ASCE Table 11.4-2, the site coefficient is

$$F_v = 1.5$$

The design response acceleration is given by ASCE Eqs. (11.4-1) and (11.4-3) as

$$\begin{aligned} S_{DS} &= 2F_a S_s/3 \\ &= 0.60 \end{aligned}$$

The design response acceleration is given by ASCE Eqs. (11.4-2) and (11.4-4) as

$$\begin{aligned} S_{D1} &= 2F_v S_1/3 \\ &= 0.50 \end{aligned}$$

For a building used as a fire station, the occupancy category is obtained from ASCE Table 1-1 as IV.

From ASCE Table 11.6-1, for a value of $S_{DS} > 0.5$, the seismic design category is D.

From ASCE Table 11.6-2, for a value of $S_{D1} > 0.2$, the seismic design category is D.

4.6 **a.** The minimum moment produced in a beam is

$$M_{min} = M_D$$
$$= 350 \text{ kip-ft}$$

The lever arm for elastic design is

$$\ell_a = d - t_s/2$$
$$= 40 - 6/2$$
$$= 37 \text{ in.}$$

The reinforcement area provided is

$$A_s = 9 \text{ in.}^2$$

The minimum reinforcement stress is given by

$$f_{min} = M_{min}/\ell_a A_s$$
$$= 350 \times 12/(37 \times 9)$$
$$= 12.61 \text{ ksi}$$

The allowable fatigue stress range is given by AASHTO Eq. (8-60) as

$$f_{f(all)} = 21 - 0.33 f_{min} + 8 \times 0.3$$
$$= 23.4 - 0.33 \times 12.61$$
$$= 19.24 \text{ ksi}$$

4.7 **c.** The prestressing force required to balance the applied load of 2 kips per foot is

$$P = WL^2/8a$$
$$= 2 \times 25^2/(8 \times 10/12)$$
$$= 187.5 \text{ kips}$$

4.8 **c.**

ASD Option

The geometrical properties of unit size of weld are

$$A = 2 \times 1 \times \ell_p$$
$$= 2 \times 24$$
$$= 48 \text{ in.}^2$$

$$S = 2 \times 1 \times \ell^2/6$$
$$= 24^2/3$$
$$= 192 \text{ in.}^3$$

The coexistent forces acting at the ends of the weld profile in the x- and y-directions are

$$f_y = R/A$$
$$= R/48$$
$$f_x = M/S$$
$$= 2R/192$$
$$= R/96$$

The resultant force at the ends of the weld profile is

$$f_R = \sqrt{f_y^2 + f_x^2}$$
$$= R\sqrt{(1/48)^2 + (1/96)^2}$$
$$= 0.0233R$$

The capacity of the weld is

$$\frac{R_n}{\Omega} = \frac{1}{\Omega} F_w A_w$$
$$= \tfrac{1}{2}\,(0.6 F_{EXX})(0.707 \tfrac{D}{16}) \text{ k/in.}$$
$$= \frac{(0.6 \times 70 \text{ ksi})(0.707 \times D)}{2 \times 16} \text{ k/in.}$$
$$= 0.928D \text{ k/in. } (D=3)$$

Equating the demand to capacity

$$f_R = \frac{R_n}{\Omega}$$

$$0.0233R = 0.928 \times 3$$
$$R = 119K$$

LRFD Option

Assuming unit size of weld, the geometrical properties of the weld are

$$A = 2 \times 1 \times \ell_p$$
$$= 2 \times 24$$
$$= 48 \text{ in.}^2$$

$$S = 2 \times 1 \times \ell^2/6$$
$$= 24^2/3$$
$$= 192 \text{ in.}^3$$

The coexistent forces acting at the ends of the weld profile in the x- and y-directions are

$$f_y = R/A$$
$$= R/48$$
$$f_x = M/S$$
$$= 2R/192$$
$$= R/96$$

The resultant force at the ends of the weld profile is

$$f_R = \sqrt{f_y^2 + f_x^2}$$
$$= R\sqrt{(1/48)^2 + (1/96)^2}$$
$$= 0.0233R$$

The capacity of the weld is

$$\phi R_n = \phi(F_w)(A_w)$$
$$= \phi(0.6F_{EXX})(0.707\frac{D}{16}) \text{ k/in.}$$
$$= 0.75 \times (0.6 \times 70 \text{ ksi})(\frac{0.707}{16}) \text{ k/in.}$$
$$= 1.39D \text{ k/in.}$$

Equating the demand to capacity,

$$0.0233R = 1.392 \times 3$$
$$R = 179 \text{ kips}$$

4.9 **a.** From IBC Table 1607.1, the specified live load for a public assembly area with movable seats is

$$w_L = 100 \text{ psf}$$

In accordance with IBC Section 1607.9.1.3, no reduction is permitted in the live load.

The dead load is given by

$$w_D = 40 + 150 \times 7.5/12$$
$$= 133.75 \text{ psf}$$

The factored loading is specified by ACI Eq. (9-2) as

$$w_u = 1.2w_D + 1.6w_L$$
$$= (1.2)(133.75) + (1.6)(100)$$
$$= 320.50 \text{ psf}$$

The moment factors given in ACI Section 8.3.3 are applicable, and the moment at the first interior support is given by

$$M_u = w_u \ell_n^2 / 10$$

where

$$\ell_n = \text{clear span} = 14.33 \text{ ft}$$

Hence,

$$M_u = 320.5 \times 14.33^2/10{,}000$$
$$= 6.58 \text{ kip-ft}$$

4.10 c. The maximum uplift on a leg is obtained with wind acting parallel to a diagonal. The wind force on the tank for wind acting parallel to a diagonal is

$$V = 34.59 \times 1.414 \times 20 \times 10/1000$$
$$= 9.78 \text{ kips}$$

Applying IBC Equation (16-14), the effective dead load is

$$W_E = 0.6 \times 24$$
$$= 14.4 \text{ kips}$$

The overturning moment on the tower is

$$M_O = VH$$
$$= 9.78 \times 35$$
$$= 342 \text{ kip-ft}$$

The restoring moment on the tower is

$$M_R = W_E \times 1.414B/2$$
$$= 14.4 \times 1.414 \times 15/2$$
$$= 153 \text{ kip-ft}$$

The maximum uplift force on one leg of the tower is

$$T = (M_O - M_R)/1.414B$$
$$= (342 - 153)/(1.414 \times 15)$$
$$= 8.94 \text{ kips}$$

4.11 d. The effective concrete flange width is given by AASHTO Section 10.38.3 as the minimum of

$$b = L/4$$
$$= 85/4$$
$$= 21.25 \text{ feet}$$

or

$$b = 12t_s$$
$$= 12 \times 8$$
$$= 96 \text{ inches}$$

or

$$b = S$$
$$= 60 \text{ inches} \ldots \text{ governs}$$

The total shear force at the interface at ultimate load is given by AASHTO Section 10.38.5.1.2 as the lesser of

$$P = A_s F_y$$
$$= 50 \times 36$$
$$= 1800 \text{ kips}$$

or

$$P = 0.85 f'_c b t_s$$
$$= 0.85 \times 3 \times 60 \times 8$$
$$= 1224 \text{ kips} \ldots \text{ governs}$$

The required number of ⅞-inch-diameter studs between midspan and support is, from AASHTO Eq. (10-60),

$$N_1 = P/\phi S_u$$
$$= 1224/(0.85 \times 30.6)$$
$$= 47 \text{ connectors}$$

4.12 c. Assuming the equivalent eccentricity does not exceed $L/6$, the maximum bearing pressure is given by

$$q_{max} = P/A + M/S$$
$$= 80/3L + 6 \times 29/3L^2$$
$$= 5 \text{ ksf}$$

Hence,

$$15L^2 - 80L - 174 = 0$$
$$L = 7 \text{ feet}$$

4.13 b. The allowable shear stress without shear reinforcement is given by BCRMS Eq. (2-20) as

$$F_v = \sqrt{f'_m}$$
$$= \sqrt{1500}$$
$$= 38.7 \text{ psi} < 50 \text{ psi}$$

The shear stress at the critical section is given by BCRMS Eq. (2-19) as

$$f_v = V/b_w d$$
$$= 14.5 \times 1000/(7.63 \times 37)$$
$$= 51.36 \text{ psi}$$
$$> 38.7 \text{ psi}$$

Hence, shear reinforcement is required to carry the total shear force. The required stirrup spacing is given by BCRMS Eq. (2-26) as

$$s = A_v F_s d/V$$
$$= 0.2 \times 24 \times 37/14.5$$
$$= 12.25 \text{ inches}$$

4.14 b. From IBC 2006 Table 16.13.5.2, the applicable site classification for this soil profile is site classification C.

4.15 b. The equivalent eccentricity of the moment acting on the pile group is

$$e = M/W$$
$$= 50/50$$
$$= 1 \text{ foot}$$

The line of piles is symmetrical with an inertia of

$$\Sigma x^2 = 2(10^2)$$
$$= 200$$

The maximum axial force occurs in the right-hand pile and, neglecting its self-weight, is given by

$$P = W/n_L + Wex/\Sigma x^2$$
$$= 50/3 + 50 \times 1 \times 10/200$$
$$= 19.2 \text{ kips}$$

4.16 b. The effective width of the concrete slab is given by AASHTO Section 10.38.3.1 as the least of

$$b = L/4$$
$$= 40 \times 12/4$$
$$= 120 \text{ inches}$$

or

$$b = 12c$$
$$= 12 \times 9$$
$$= 108 \text{ inches}$$

or

$$b = S$$
$$= 6 \times 12$$
$$= 72 \text{ inches} \ldots \text{ governs}$$

4.17 c. The impact fraction is given in AASHTO section 3.8.2.1 as

$$I = 50/(L + 125)$$
$$= 50/(40 + 125)$$
$$= 0.303$$
$$> 0.30$$

Hence, the maximum value of $I = 0.30$ governs.

4.18 c. The distribution of wheel loads to each longitudinal girder for a girder spacing of $S < 14$ feet is obtained from AASHTO Table 3.23.1 as

$$G = S/D$$
$$= 6/5.5$$
$$= 1.09$$

4.19 a. The maximum live load moment, not including impact, at midspan produced by loading one design lane with HS20-44 standard truck loading is obtained from AASHTO Appendix A as

$$M_L = 449.8 \text{ kip-ft}$$
$$I = 0.30$$
$$G = 1.09$$

The maximum live load moment in an interior beam is

$$M = M_L(1 + I)G/2$$
$$= 449.8(1 + 0.30) \times 1.09/2$$
$$= 319 \text{ kip-ft}$$

4.20 c. The end reaction, not including impact, produced by loading one design lane with HS20-44 standard truck loading is obtained from AASHTO Appendix A as

$$V_L = 55.2 \text{ kips}$$
$$I = 0.30$$
$$G = 1.09$$

The end reaction in an interior beam is

$$V = V_L(1 + I)G/2$$
$$= 55.2(1 + 0.30) \times 1.09/2$$
$$= 39 \text{ kips}$$

4.21 c. From ASCE 7-05 Tables 9.4.1.2a and 9.4.1.2b and IBC 2006 Tables 1613.5.3(1) and 1613.5.3(2), the site coefficients are determined as

$$F_a = 1.00$$
$$F_v = 1.50$$

4.22 d. From IBC 2006 Tables 1613.5.3(1) and (2) and Equations 16-37, 16-38, 16-39, and 16-40, the design spectral response accelerations at short periods and at a period of 1 second are

$$S_{DS} = 2F_a S_S/3$$
$$= 2 \times 1.00 \times 1.25/3$$
$$= 0.83g$$

and
$$S_{D1} = 2F_v S_1/3$$
$$= 2 \times 1.30 \times 0.50/3$$
$$= 0.43g$$

4.23 a. The approximate fundamental period for a wood-framed structure is given by ASCE Equation (9.5.5.3.2-1) as

$$T_a = 0.020(h_n)^{0.75}$$

where
h_n = roof height
= 15 ft

Thus, the fundamental period is

$$T_a = 0.020(15)^{0.75}$$
$$= 0.15 \text{ second}$$

4.24 c.

ASD Option

The relevant properties of the W14 × 68 are

$$A_g = 20 \text{ in.}^2$$
$$b_f = 10 \text{ in.}$$
$$t_f = 0.72 \text{ in.}$$

Both flanges are welded by transverse welds to gusset plates, and the area of directly connected elements is

$$A_e = 2b_f t_f$$
$$= 2 \times 10 \times 0.72$$
$$= 14.4 \text{ in.}^2$$

The tensile capacity for the yielding condition is (Equation 2-1)

$$P_t = F_y A_g / 1.67$$
$$= 50 \times 20 / 1.67$$
$$= 598.8 \text{ kips}$$

The tensile capacity for the fracture condition is (Equation D2-2)

$$P_t = F_u A_e / 2.0$$
$$= 65 \times 14.4 / 2.0$$
$$= 468 \text{ kips ... governs}$$

LRFD Option

The relevant properties of the W14 × 68 are

$$A_g = 20 \text{ in.}^2$$
$$b_f = 10 \text{ in.}$$
$$t_f = 0.72 \text{ in.}$$

Both flanges are welded by transverse welds to gusset plates, and the area of directly connected elements is

$$A_e = 2b_f t_f$$
$$= 2 \times 10 \times 0.72$$
$$= 14.4 \text{ in.}^2$$

The design axial strength for the yielding condition is (Equation D2-1)

$$\phi_t P_n = 0.9 F_y A_g$$
$$= 0.9 \times 50 \times 20$$
$$= 900 \text{ kips}$$

The design axial strength for the fracture condition is (Equation D2-2)

$$\phi_t P_n = 0.75 F_u A_e$$
$$= 0.75 \times 65 \times 14.4$$
$$= 702 \text{ kips ... governs}$$

4.25 c. For a standard occupancy structure, IBC Tables 1613.5.6(1) and (2) give a seismic use group designation of I.

For a design spectral response acceleration at short periods of

$$S_{DS} = 0.80g$$
$$> 0.50g$$

IBC Table 1613.5.6(1) gives the seismic design category as D.

For a design spectral response acceleration at long periods of

$$S_{D1} = 0.50g$$
$$> 0.20g$$

IBC Table 1613.5.6(2) gives the seismic design category as D.

Hence, the seismic design category is D.

4.26 d. Concrete cover to the tension reinforcement is

$$c_c = 1.5 + 0.5$$
$$= 2 \text{ inches}$$

Stress in the reinforcement at service load may be taken as

$$f_s = 2 f_y / 3$$
$$= 2 \times 60 / 3$$
$$= 40 \text{ ksi}$$

The maximum allowable spacing of tension reinforcement is given by ACI Equation (10-4) as

$$s = 15 (40,000 / f_s) - 2.5 c_c$$
$$= 15 - 2.5 \times 2$$
$$= 10 \text{ inches} < 12 (40,000 / f_s) = 12 \text{ inches}$$

4.27 c.

$$\text{Void ratio, } e = \frac{n}{1-n} = \frac{0.4}{1-0.4} = 0.667$$

$$\gamma = \frac{G_s\gamma_w + wG_s\gamma_w}{1+e} = \frac{G_s\gamma_w(1+w)}{1+e}$$

$$17.8 = \frac{(G_s)(9.81)(1+0.1)}{1+0.667}$$

$$G_s = 2.75$$

4.28 c.

$$\text{Dry unit weight in field, } \gamma_d = \frac{\gamma}{1+w} = \frac{122}{1+0.12} = 108.92 \text{ lb/ft}^3$$

$$\text{Relative compaction, } R = \frac{\gamma_d}{\gamma_{d(\max)}} = \frac{108.92}{118} = 0.92 = 92\%$$

4.29 b.

$$\text{Dry unit weight, } \gamma_d = \frac{G_s\gamma_w}{1+e} = \frac{(2.65)(9.81)}{1+0.5} = 17.33 \text{ kN/m}^3$$

$$\text{Saturated unit weight, } \gamma_{sat} = \frac{(G_s+e)\gamma_w}{1+e} = \frac{(2.65+0.5)(9.81)}{1+0.5}$$
$$= 20.6 \text{ kN/m}^3$$

Effective stress at $A = (\gamma_d)(2) + (\gamma_{sat})(3) - (\gamma_w)(3)$
$$= (17.33)(2) + (20.6)(3) - (9.81)(3) = 67.03 \text{ kN/m}^2$$

4.30 d. The friction angle can be calculated from the given shear stress (τ) and normal stress (σ) values from the following equation:

$$\tan\phi = \frac{\tau}{\sigma} = \frac{11}{15}$$

$$\phi = 36.25°$$

4.31 b. The Rankine active pressure coefficient can be calculated as

$$K_a = \tan^2\left(45 - \frac{\phi}{2}\right) = \tan^2\left(45 - \frac{36}{2}\right) = 0.26$$

Thus, the intensity of active pressure is

$$\sigma_a = \gamma z K_a = (110)(10)(0.26) = 286 \text{ lb/ft}^2$$

4.32 c. The Rankine active pressure coefficient can be calculated as

$$K_a = \tan^2\left(45 - \frac{\phi}{2}\right) = \tan^2\left(45 - \frac{35}{2}\right) = 0.271$$

Refer to Exhibit 4.32b.

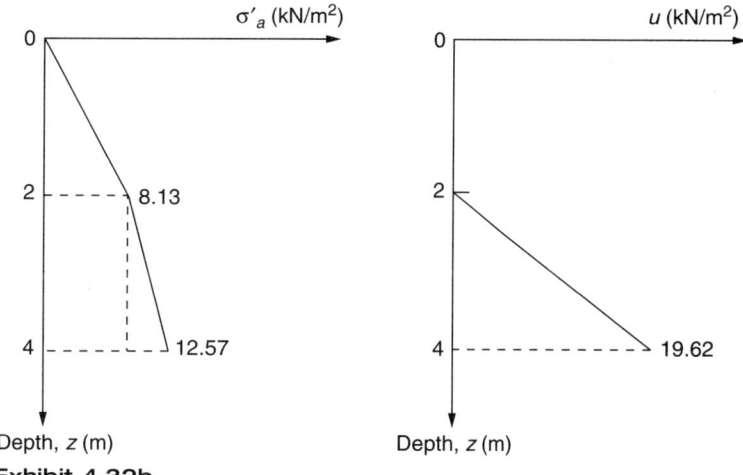

Exhibit 4.32b

At $z = 0$: $\quad \sigma'_a = 0$; pore water pressure $u = 0$
At $z = 2$ m: $\quad \sigma'_a = (2 \times 15)(K_a) = 8.13$ kN/m^2; $\quad u = 0$
At $z = 4$ m: $\quad \sigma'_a = [(2 \times 15) + (2)(18 - 9.81)](0.271) = 12.57$ kN/m^2

$$u = (2)(9.81) = 19.62 \text{ kN/m}^2$$

P_a = area of the two pressure diagrams
$$= \left(\frac{1}{2}\right)(2)(8.13) + \left(\frac{8.13 + 12.57}{2}\right)(2) + \left(\frac{1}{2}\right)(2)(19.62)$$
$$= 8.13 + 20.7 + 19.62 = 48.45 \text{ kN/m}$$

4.33 d. For square foundations,
$$q_u = 1.3cN_c + qN_q + 0.4\gamma BN_\gamma$$
$$= (1.3)(400)(31.61) + (115 \times 3)(17.81) + (0.4)(115)(3)(13.7)$$
$$= 16{,}437.2 + 6144.45 + 1890.6 = 24{,}472.25 \text{ lb/ft}^2$$

The ultimate load $Q = q_u B^2 = (24{,}472.25)(9) = 220{,}250.25$ lb ≈ 220 kip.

4.34 d.

Effective width $B' = B - 2e = 4 - (2)(0.4) = 3.2$ ft.
$$q_u = qN_q + \frac{1}{2}\gamma B'N_\gamma$$
$$= (120 \times 3)(33.3) + \left(\frac{1}{2}\right)(120)(3.2)(48.03)$$
$$= 11{,}988 + 9221.76 = 21{,}209.76 \text{ lb/ft}^2$$
$Q_u = (q_u)(B' \times 1) = (21{,}209.76)(3.2 \times 1) = 67{,}871.23$ lb/ft ≈ 67.9 kip/ft

4.35 d.
$$Q_u = Q_p + Q_s = (c_u N_c)(A) + \alpha c_u L p$$

N = bearing capacity factor = 9
A = area of pile cross section
L = pile length

p = perimeter of pile

$$Q_u = (1600 \times 9)\left(\frac{18}{12} \times \frac{18}{12}\right) + (0.5)(1600)(30)\left(4 \times \frac{18}{12}\right)$$
$$= 32,400 + 144,000 = 176,400 \text{ lb} \approx 176.4 \text{ kip}$$

4.36 b.

$$Q_p = A_p q N_q^* \leq 1000 A_p N_q^* \tan \phi$$

$$A_p = \frac{18}{12} \times \frac{18}{12} = 2.25 \text{ ft}$$

$$q = (115)(60) = 6900 \text{ lb/ft}^2$$

$$N_q^* = 120$$

$$A_p q N_q^* = (2.25)(6900)(120) = 1,863,000 \text{ lb} \approx 1863 \text{ kip}$$

However,

$$1000 A_p N_q^* \tan \phi = (1000)(2.25)(120)(\tan 35) = 189,056 \text{ lb} \approx 189 \text{ kip}$$

So, $Q_p = 189$ kip.

4.37 d. The radius of curvature of the centerline of the inside lane is 400 − 0.5(3.6) = 398.2 m. The middle ordinate from the centerline of the inside lane to the obstruction to vision is 0.5(3.6) + 2.4 + 2 = 6.2 m. From Exhibit 3-53 of the AASHTO Green Book (AASHTO, 2004, p. 226), the sight distance is approximately 145 m.

4.38 a. The elevation of the two-lane highway is increasing throughout the vertical curve, which extends beyond the location of the grade separation structure, so the minimum clearance occurs at the far side of the structure. Also, the grade separation structure has a slope that is less than the cross slope of the two-lane highway, so the critical clearance will occur at the centerline of the two-lane highway. The minimum elevation of the bottom of the structure, however, occurs at the right edge of the traveled way of the road. The distance from the PVC to the critical point is 1050 m − 860 m + 0.5(14 m) = 197 m. The elevation of the centerline of the highway at this point is given by

$$E_x = E_{pvc} + g_1 x - ax^2$$
$$= 152.000 + 0.030(197) - [(0.030 - 0.005)(197^2)]/[2(300)]$$
$$= 156.293$$

The minimum elevation of the grade separation structure at the centerline of the two-lane highway is 156.293 + 5.0 = 161.293, and the elevation of the bottom of the grade separation structure at the right edge of the traveled way of the highway is 161.293 − 0.01(3.6) = 161.257.

4.39 c. Volumes between the stations for which cross section areas are given may be calculated by Eq. 11.154b in *Civil Engineering PE License Review*:

Station 50 + 00 to 50 + 40: $V = (120.25 + 60.75)(40/54) = 134.1$ yd^3

Station 50 + 40 to 51 + 20: $V = (60.75 + 15.87)(80/54) = 113.5$ yd^3

The total excavation is 134.1 + 113.5 = 247.6 yd^3.

4.40 c. According to AASHTO Exhibit 6-6 (AASHTO, 2004, p. 426), the minimum clear roadway width for a new or reconstructed bridge for a roadway with a design volume of 1700 vehicles per day is the traveled way plus 4 feet (with an exception, which does not apply, for bridges longer than 100 ft). The width of the traveled way is (2)(11 ft) = 22 ft. The minimum required width for the bridge is 22 ft + 2(4 ft) = 30 ft.

SOLUTIONS TO CHAPTER 5 PROBLEMS

5.1 d. Space-mean speed = 5/[(1/57.6) + (1/62.4) + (1/62.1) + (1/74.4) + (1/49.6)] = 60.2 mph. Time-mean speed = (57.6 + 62.4 + 62.1 + 74.4 + 49.6)/5 = 61.2 mph.

5.2 d. Warrant 1 may be met in three ways: (1) there are a total of eight hours for which both the major street and the minor street volumes exceed the values in the 100 percent column for Condition A in Table 4C-1 of the MUTCD (FHWA, 2003, p. 4C-3), (2) there are a total of eight hours for which both the major street volumes and minor street volumes exceed the values in the 100 percent column for Condition B, or (3) there are a total of eight hours for which volumes exceed those in the 80 percent column for both Condition A and Condition B (the hours exceeded for Condition A do not have to be the same as for Condition B).

For the situation given, the volumes in the 100 percent column for Condition A are exceeded during the following hours: 0700–0800, 0800–0900, 1500–1600, and 1600–1700. This is a total of four hours, so criterion 1 is not met.

The volumes in the 100 percent column for Condition B are exceeded for the following hours: 0700–0800 and 1600–1700. This is a total of two hours. Criterion 2 is not met.

The volumes in the 80 percent column for Condition A are exceeded during the following hours: 0600–0700, 0700–0800, 0800–0900, 0900–1000, 1200–1300, 1300–1400, 1400–1500, 1500–1600, and 1600–1700. This is a total of nine hours. The volumes in the 80 percent column for Condition B are exceeded during the following hours: 0600–0700, 0700–0800, 0800–0900, 1500–1600, 1600–1700. This is a total of five hours. Since the volumes in the 80 percent column are not exceeded for eight or more hours for *both* conditions, criterion 3 is also not met. Therefore, warrant 1 is not satisfied.

Warrant 2 is satisfied if volumes for at least four hours plot above the applicable line in Fig. 4C-1 of the MUTCD. Hours that plot above the line are 0700–0800 and 1600–1700. This is only two hours, so warrant 2 is not satisfied.

Warrant 3 is satisfied if volumes for at least one hour fall above the line in Fig. 4-C3 of the MUTCD. No hours do, so warrant 3 is not satisfied.

Consequently, none of the warrants is satisfied.

5.3 c. Intersection sight distance at stop-controlled intersections is described by departure sight triangles shown in Exhibit 9-50 of the Green Book (AASHTO, 2004, p. 652). For case B, intersections with stop control on the minor roadway, three cases must be considered: left turns from the minor road, right turns from the minor road, and crossing maneuvers by vehicles from the minor road approach. Design intersection sight distances for these cases are given by Exhibits 9-55 and 9-59 of the Green Book. For a design speed of 100 km/h, Exhibit 9-55 gives a design intersection sight distance of 210 m for left turns from the minor road, and Exhibit 9-59 gives a value of 185 m for right turns from the minor road and crossing maneuvers. Since all three types of maneuvers must be accommodated, the required design intersection sight distance is 210 m.

5.4 d. Free-flow speed is estimated by

$$\text{FFS} = \text{BFFS} - f_{\text{LW}} - f_{\text{LC}} - f_N - f_{\text{ID}}$$

The base free-flow speed, BFFS, is 70 mph for urban freeways. From Exhibit 23-4 of the HCM (TRB, 2000), there is no adjustment for lane width, since it is 12 ft. From Exhibit 23-5, there is no adjustment for right shoulder clearance because the 10 ft shoulder provides a clearance of more than 6 ft. From Exhibit 23-6, the adjustment f_N for the number of lanes is 3.0 mph for three lanes in one direction. The interchange density is 7 interchanges/6 mi = 1.167 interchanges/mi. In Exhibit 23-7, this falls between 1.00 miles and 1.25 miles. Interpolating, the adjustment for interchange density, f_{ID}, is

$$f_{\text{ID}} = 2.5 + [(1.167 - 1.000)/(1.25 - 1.00)](3.7 - 2.5) = 3.3 \text{ mph}$$
$$\text{FFS} = 70 - 0.0 - 0.0 - 3.0 - 3.3 = 63.7 \text{ mph}$$

5.5 d. The total flow entering the ramp influence area is given by $v_{R12} = v_R + v_{12}$. In turn, v_{12} is given by $v_{12} = P_{\text{FM}} v_F$. From Exhibit 25-5 of the HCM, P_{FM} for a freeway with four lanes in one direction is

$$P_{\text{FM}} = 0.2178 - 0.000125 v_R + 0.05887 L_A/S_{\text{FR}}$$
$$= 0.2178 - 0.000125(700) + 0.05887(200/60) = 0.3265$$
$$v_{12} = (0.3265)(7200) = 2351 \text{ pc/h}$$
$$v_{R12} = 2351 + 700 = 3051 \text{ pc/h}$$

5.6 b. $T_{ij} = 1000\{(1200)(30)/[(400)(12) + (1200)(30) + (700)(16) + (500)(22)]\} = 571.4$

5.7 d. Portions of the mass diagram that have a positive slope indicate a surplus of cut over fill, and those with a negative slope indicate that more fill is required than can be supplied by the cut (if any) in that section. The longest distance material will be hauled is the limit of economic haul. Thus, any material represented by a negatively sloped portion of the mass diagram that is outside the portions of the loops cut off by the limit of economic haul lines must be borrowed. In the mass diagram in question, there are three such areas, as indicated in Exhibit 5.7b.

These total $10,000 + 15,000 + 5000 = 30,000$ yd^3.

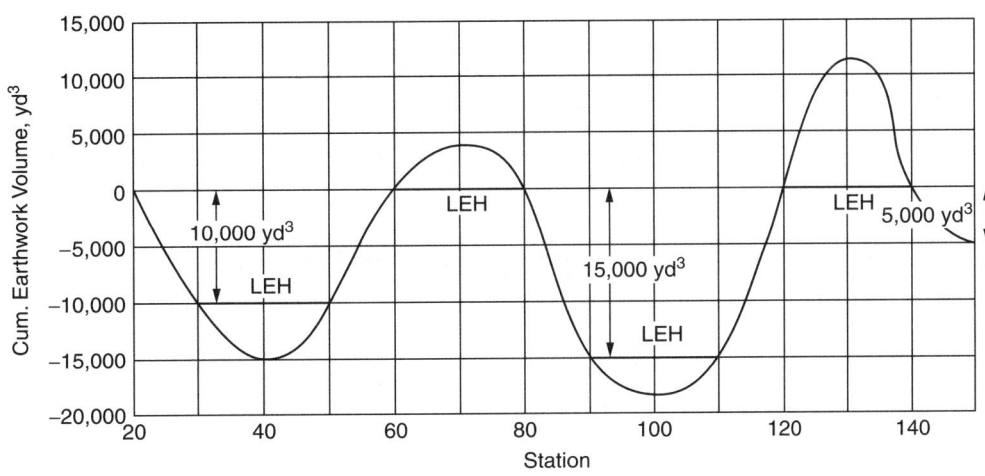

Exhibit 5.7b

5.8 a. The structural number of the asphalt surface is determined using the nomograph presented as Fig. 3.1 in the AASHTO *Guide for Design of Pavement Structures* (1993). Enter the nomograph with the following values: $R = 90$, $S_o = 0.45$, $W_{18} = 3$, $M_R = 24$ (the elastic modulus of the next lower layer, the granular base, divided by 1000), and $\Delta PSI = 1.7$. The resulting design structural number, SN, is approximately 2.7.

5.9 c. The minimum length of superelevation runoff is given by Eq. 3.25 in the Green Book (AASHTO, 2004, p. 177) as

$$L_r = [(wn_1)e_d/\Delta](b_w)$$

In this case, $w = 3.6$ m (given), $n_1 = 2$ (assuming rotation about the centerline), $e_d = 5.3\%$ (given), $\Delta = 0.44$ (Exhibit 3-30), and $b_w = 0.75$ (Exhibit 3-31). The minimum length of superelevation runoff is

$$L_r = [(3.6)(2)(5.3)/(0.44)](0.75) = 65 \text{ m}$$

5.10 d. Roadway elevation at station $18 + 94 = 328.5 + 2.2 = 330.7$ ft

Tangent elevation at station $18 + 94 = 325.00 + (-1.5 \text{ sta})(-1.5\%) = 327.25$ ft

Vertical curve offset at station $18 + 94 = 330.7 - 327.25 = 3.45$ ft

At station $18 + 94$, $x = L/2 - 1.5$ sta

The difference in grade $A = g_2 - g_1 = 2.3 - (-1.5) = 3.8\%$

The offset = $(A/2L)x^2$. Therefore,
$$(3.8/2L)(L/2 - 1.5)^2 = 3.45$$
$$3.8(L/2 - 1.5)^2 = 6.9L$$
$$0.95L^2 - 12.6L + 8.55 = 0$$

Solving the quadratic equation,

$$L = \left\{12.6 \pm \sqrt{(12.6)^2 - 4(0.95)(8.55)}\right\}/[2(0.95)] = (12.6 \pm 11.24)/1.9$$

Therefore, $L = 12.49$ sta or $L = 0.72$ sta. Since $L = 0.72$ sta would place the point with the critical clearance outside the vertical curve, the minimum length of the curve is 12.49 sta = 1249 ft.

Note: The solution of this problem can be simplified by use of a formula in Banks (2002).

5.11 d. Braking distances for the initial and final speeds of the vehicle are

$$S_{bi} = v_i^2/[2g(f+G)] \quad \text{and} \quad S_{bf} = v_f^2/[2g(f+G)]$$

The length of the skid represents the differences between the two stopping distances, so

$$d = v_i^2/[2g(f+G)] - v_f^2/[2g(f+G)]$$

Solving for the initial speed,

$$v_i^2 = \sqrt{2g(f+G)d + v_f^2}$$
$$= \sqrt{2(32.2)(0.30 + 0.015)(150) + [20(5280/3600)]^2}$$
$$= 62.47 \text{ ft/s}$$

$$62.47(3600/5280) = 42.3 \text{ mph}$$

5.12 b. Total accidents = 12 + 35 + 15 = 62.

Annual vehicle miles of travel = $[(10,000)(1.5) + (14,000)(3.0) + (17,000)(1.8)](365) = 31,974,000$

Accident rate = $1,000,000[(62)/(31,974,000)]$
= 1.94 accidents per million vehicle-miles

5.13 c. Accidents per year = $(15 + 18 + 12)/3 = 15$

Accidents prevented = $[(15)(0.32)(7500)]/6000 = 6.0$

5.14 c. Assuming left-turn movements can be paired with one another and/or the parallel through movement, use critical approach v/s:

NBLT + SB	0.17 + 0.31 = 0.48
SBLT + NB	0.15 + 0.34 = 0.49 *critical*
EB	0.27
WB	0.31 *critical*

The optimum cycle length is given by

$$C = \frac{1.5L+5}{1-\Sigma Y_i} = \frac{1.5(3\times 3\text{s})+5}{1-(0.15+0.34+0.31)} = 92.5\,\text{s}$$

5.15 b. Cycle time = 10 + 35 + 25 + 40 + 10 + 42 + 26 + 33 = 221 min

Minimum number of vehicles = cycle time/headway = 221/15 = 14.7. Round up to 15 vehicles.

5.16 a. The algebraic difference in grade is 2.0 − (−4.0) = 6%. From Exhibit 3-71 of the AASHTO Green Book (2004, p. 271), the minimum vertical curve length required to provide stopping sight distance for a design speed of 60 mph is approximately 900 ft.

5.17 c. Guidance given in MUTCD, Section 2B.13 (FHWA, 2003, p. 2B-10), states that "when a speed limit is posted, it should be within 10 km/h or 5 mph of the 85th-percentile speed of freely-flowing traffic." From the table, the 85th-percentile speed is between 52.6 mph and 57.5 mph; therefore, the appropriate speed limit is 55 mph.

5.18 b. The deflection angle $\Delta = 75° − 30° = 45°$. The distance from the PI to the PC is given by $T = R \tan(\Delta/2) = 600 \tan(45°/2) = 248.528$ m = 0 + 248.582 stations. The PC station is (1 + 127) − (0 + 248.582) = 0 + 878.418.

5.19 d. Space-hours of demand during the peak four-hour period is $D = (50)(4.0) + (375)(0.5) = 387.5$. Since all spaces can be legally parked on for the entire four-hour period, the supply in space-hours is $S = fNt = 0.85(4)N = 387.5$. Solving for N, the required number of spaces is 113.97, which is rounded up to 114 spaces.

5.20 b. The definition of the monthly expansion factor is $F_m = \text{AADT}/\text{ADT}_m$. Solving for annual average daily traffic, $\text{AADT} = F_m \text{ADT}_m = (0.95)(2500) = 2375$.

5.21 c. Assuming that the distribution of trips generated is normal, the probability that the number of trips is greater than one standard deviation above the mean is approximately 0.15. The required estimate of the number of trips generated is $(120{,}000/1000)(4.24 + 2.23) = 776.4$ trips.

5.22 b. The average pedestrian delay is $d_p = 0.5(C − g)^2/C = 0.5(90 − 18)^2/90 = 28.8$ s. From Exhibit 18-9 of the HCM (TRB, 2000, p. 18-8), the level of service is C.

5.23 c. The ideal offset is

$$O_I = x/V = (90 + 140 + 120)/[50(1000/3600)] = 25.2\,\text{s}$$

5.24 b. A coupling index may be used to determine whether signal coordination should be tried. The index is given by $I = V/x$. For the four streets in question, the index is as follows: Elm Street, $I = 700/600 = 1.17$; Ash Street, $I = 450/1200 = 0.38$; Poplar Street, $I = 400/900 = 0.44$; and Walnut Street, $I = 800/450 = 1.78$. Coordination should be considered if the index exceeds 0.5. This is true for Elm Street and Walnut Street.

5.25 a. The distance required to stop is the sum of the distance traveled during the reaction time and the braking distance. This is

$$d_s = 1.47Vt + \frac{V^2}{30[(a/32.2) \pm G]} = 1.47(30)(2.5) + \frac{30^2}{30[(11.2/32.2) + .015]}$$
$$= 192.9 \text{ ft}$$

5.26 c. Curve radius is related to design speed, superelevation rate, and friction factor by

$$R = v^2/127(e + f_s)$$

Solving for f_s, $f_s = (v^2/127R) - e = 100^2/[(127)(500)] - 0.09 = 0.067$.

5.27 c. Degree of saturation, $S = \dfrac{wG_s}{e}$

$$S = \frac{\left(\frac{15}{100}\right)(2.68)}{0.8} = 0.5 = 50\%$$

5.28 a. Moisture content, $w = \left(\dfrac{432 - 320}{320}\right)(100) = 35\%$

$$\text{Liquidity index} = \frac{w - PL}{LL - PL} = \frac{35 - 21}{48 - 21} = 0.52$$

5.29 c.

$$k = C\frac{e^n}{1+e}$$

$$\frac{k_1}{k_2} = \frac{\dfrac{e_1^n}{1+e_1}}{\dfrac{e_2^n}{1+e_2}}$$

$$\frac{0.6 \times 10^{-7}}{1.519 \times 10^{-7}} = \left(\frac{1.2}{1.52}\right)^n \left(\frac{2.52}{2.2}\right)$$

$$n = 4.5$$

$$k = 0.6 \times 10^{-7} = C\left(\frac{1.2^{4.5}}{1+1.2}\right)$$

$C = 0.581 \times 10^{-7}$ cm/s. So,

$$k = (0.581 \times 10^{-7}) \frac{e^{4.5}}{1+e}$$

$$k_{1.4} = (0.581 \times 10^{-7}) \left(\frac{1.4^{4.5}}{1+1.4} \right) = 1.1 \times 10^{-7} \text{ cm/s}$$

5.30 c. The hydraulic gradient $i = \dfrac{12 \text{ ft}}{\left(\dfrac{150 \text{ ft}}{\cos 8} \right)} = 0.0792$

The rate of flow, $q = kiA = \left(\dfrac{0.03}{12} \text{ ft/s} \right)(0.0792)(10 \cos 8)$

$= 1.96 \times 10^{-3} \text{ ft}^3\text{/s/ft}$

5.31 d. Total length cored = 2.5 ft = 30 in.

Recovery length = (2)(6) + (2)(8) = 28 in.

Recovery ratio = $\dfrac{28}{30} \times 100 = 93.3\%$

5.32 a. Refer to the AASHTO soil classification table in any textbook.

Percent passing No. 200 sieve F_{200} = 50 (greater than 36)
Liquid limit LL = 31 (less than 40)
Plasticity index $PI = LL - PI = 31 - 12 = 19$ (greater than 11)

So the soil classification is *A*-6.

Group index:

$GI = (F_{200} - 35)[0.2 + 0.005(LL - 40)] + 0.01(F_{200} - 15)(PI - 10)$

$= (52 - 35)[0.2 + 0.005(31 - 40)] + 0.01(52 - 15)(19 - 10) = 5.965 \approx 6$

5.33 c. Given that the overflow rate = (flow)/(area), the area of the tank is

area = flow/overflow rate
= (0.2 m³/s)/(20 m/day)
= 864 m²

The volume of the tank is the tank area multiplied by the tank depth:

volume = area × depth = 864 m² × 2.5 m = 2160 m³

Finally, the detention time θ is

θ = volume/flow = (2160 m³)/(0.2 m³/s) = 180 minutes

5.34 d. If the flow is steady and uniform, the velocity must be constant throughout the entire channel.

5.35 a. A semicircle has the most efficient hydraulic cross section because this shape maximizes the hydraulic radius.

5.36 c. By definition, flow changes from supercritical to subcritical across a hydraulic jump.

5.37 a. The intensity corresponds to the time of concentration (t_c) to Inlet Y. By inspection, t_c for Inlet Y is the time to travel within Catchment X plus the time of travel (t_t) in the pipe between Inlet X and Inlet Y.

The time of travel is the pipe length divided by the pipe velocity. Pipe velocity can be estimated by Manning's Equation, assuming full pipe flow.

$$V = 1.486 \times n^{-1} \times (R_h)^{2/3} \times (S_o)^{1/2}$$

where
R_h for an 18-inch circular cross section is $D/4$, or (18 inches)/4, or 0.375 ft
$n = 0.013$ for concrete pipe
$S_o = 0.02$

This yields a V of 8.41 ft/s and a t_t of (1000 ft)/(8.4 ft/s) = 120 min.

Given this t_c (120 min + 10 min = 130 min), the intensity from the provided intensity-duration curve is estimated to be around 1 inch per hour.

5.38 c. The full-pipe velocity is given by Q/A, or 1 cfs/($\pi \times (0.5 \text{ ft})^2$), or 1.3 fps. By definition or inspection of a hydraulic elements chart, the half-full velocity equals the full-pipe velocity, and thus the half-full flow is half of the full-pipe flow.

5.39 c. Fire flow requirements for a residential neighborhood of this size (roughly $\frac{1}{2}$-acre lots) can be assumed to be 500 gpm.

The average day demand can be assumed to be anywhere between 50 gal/capita · day and 100 gal/capita · day. Assuming 2–4 residents per home (i.e., 200–400 residents in the subdivision) yields a possible range of average day demands of 10,000 gal/day to 40,000 gal/day, or 7 gpm to 28 gpm.

Therefore, the ratio between fire flow to average day demand ranges from 18 to 70.

5.40 d. The peak of the hydrograph will occur earlier because development typically reduces the time of concentration for a catchment.

The peak of the hydrograph will be greater following development because the peak flow is increased. The peak flow is increased due to an increase in impervious area.

Not only will the peak flow be greater, but the total volume of runoff will also be greater following development due to decreased infiltration. Recall that the area under a hydrograph represents the volume of runoff.

SOLUTIONS TO CHAPTER 6 PROBLEMS

6.1 c. The client can only recover the loss incurred by the default, which is the difference between the low bid and the second lowest bid: $20,000.

6.2 b. Total cost = $50,000 + $5,000 (O&P) = $55,000.

Unit cost = $55,000 / 200 = $275 / ft. Note that the unit price must include the contractor's profit/markup.

6.3 c. The Measured Mile (MM) method compares the productivity of the impacted period of the project by a negative condition to the productivity of similar work under normal, unimpacted conditions.

The total labor cost during the impacted conditions is determined by multiplying the total daily labor cost by the impacted period:

$$\text{Total Daily Labor Cost} = (\$12/\text{hr} + \$8/\text{hr}) \times 8 \text{ hrs/day} \times 1.40 = \$224/\text{day}$$
$$\text{Total Impacted Labor Cost} = \$224/\text{day} \times 50 \text{ days} = \$11,200$$

The daily production rate under normal conditions equals 100 bricks/hr × 8 hrs/day for a total of 800 bricks/day. This rate would yield a total of 40 days for the installation of the remaining 32,000 bricks. Under normal conditions, the estimated total cost would be 40 days × $224/day for a total of $8,960. Therefore, the contractor's compensable delay is the difference of the impacted labor cost and unimpacted labor cost:

$$\text{Compensable delay} = \$11,200 - \$8,960 = \$2,240$$

6.4 b. Area Calculations

c = height of fill
b = width of roadway
s = side slope in the form s:1

Area = $c(b + sc)$ for a level cross section
$A_1 = (5 \text{ ft})[32 \text{ ft} + (2)(5 \text{ ft})] = 210 \text{ ft}^2$
$A_2 = (7 \text{ ft})[32 \text{ ft} + (2)(7 \text{ ft})] = 322 \text{ ft}^2$

Alternate way to find area 1:

Area of a trapezoid = $[(w_a + w_b) / 2] \times h$
$w_a = 32 \text{ ft}$
$w_b = 2(5 \text{ ft}) + 32 \text{ ft} + 2(5 \text{ ft}) = 52 \text{ ft}$
$A_1 = [(32 \text{ ft} + 52 \text{ ft}) / 2] \times (5 \text{ ft}) = 210 \text{ ft}^2$

Volume Calculation

A_1 and A_2 are the areas (ft^2) for the respective end areas.

L is the length (ft) between end areas.

$$\text{Volume (yd}^3) = \frac{(A_1+A_2)(L)}{2} \frac{1 \text{ yd}^3}{27 \text{ ft}^3}$$

$$= \frac{(210 \text{ ft}^2 + 322 \text{ ft}^2)(80\text{ft})}{2} \frac{1 \text{ yd}^3}{27 \text{ ft}^3}$$

$$= 788 \text{ yd}^3$$

6.5 c. Area = 40' × 40' = 1600 sq ft
Avg fill = (2.5' + 3.2' + 2.8' + 3.5')/4 = 3.0'
Volume = (1600 × 3)/27 ft³/yd³ = 177.8 yd³

6.6 a. Area Calculations

c = height of fill
b = width of roadway
s = side slope in the form s:1

Area = $c(b + sc)$ for a level cross section
A_1 = (5 ft) [32 ft + (2) (5 ft)] = 210 ft²
A_2 = (7 ft) [32 ft + (2) (7 ft)] = 322 ft²

$c_{average}$ = (5 ft + 7 ft) / 2 = 6 ft
A_m = (6 ft) [32 ft + (2) (6 ft)] = 264 ft²

Alternate way to find area 1:

Area of a trapezoid = $[(w_a + w_b) / 2] \times h$
w_a = 32 ft
w_b = 2(5 ft) + 32 ft + 2(5 ft) = 52 ft

A_1 = [(32 ft + 52 ft) / 2] × (5 ft) = 210 ft²

Volume Calculation
A_1 and A_2 are the areas (ft²) for the respective end areas.
A_m is the area obtained by averaging the dimensions of the end areas.
L is the length (ft) between end areas.

$$\text{Volume (yd}^3) = \frac{(A_1 + 4A_m + A_2)(L)}{6} \frac{1 \text{ yd}^3}{27 \text{ ft}^3}$$

$$= \frac{[210 \text{ ft}^2 + 4(264 \text{ ft}^2) + 322 \text{ ft}^2](80\text{ft})}{6} \frac{1 \text{ yd}^3}{27 \text{ ft}^3}$$

$$= 784 \text{ yd}^3$$

6.7 a.

$V = (1/2) BHL$ (triangular spoil bank)
V = volume
B = width
H = height
L = length

$$20' 5'' = 20' + 5''(1'/12'') = 20.4166667 \text{ ft}$$
$$45' 9'' = 45' + 9''(1'/12'') = 45.75 \text{ ft}$$

$$V_1 = (1/2)(20.4166667 \text{ ft})(10 \text{ ft})(45.75 \text{ ft}) = 4670.3 \text{ ft}^3$$
$$V_1 = [4670.3 \text{ ft}^3] [(1 \text{ yd}^3)/(27 \text{ ft}^3)] = 172.97 \text{ yd}^3 = 173 \text{ yd}^3$$

$$30' 11'' = 30' + 11'' (1'/12'') = 30.9166667 \text{ ft}$$

$$V = (1/3) (\Pi D^2 / 4)H \quad \text{(conical spoil pile)}$$
$$V = \text{volume}$$
$$D = \text{base diameter}$$
$$H = \text{height}$$

$$V_2 = (1/3) (\Pi) (30.9166667 \text{ ft})^2 / 4) (15 \text{ ft}) = 3753.6 \text{ ft}^3$$
$$V_2 = [3753.6 \text{ ft}^3] [(1 \text{ yd}^3)/(27 \text{ ft}^3)] = 139.02 \text{ yd}^3 = 139 \text{ yd}^3$$

$$V_{\text{total}} = V_1 + V_2 = 173 \text{ yd}^3 + 139 \text{ yd}^3 = 312 \text{ yd}^3$$

6.8 d. First, determine the amount of bank measure needed:
$$5000 \text{ yd}^3 / 0.75 = 6667 \text{ yd}^3$$

Second, determine the amount of loose soil:
$$6667 \text{ yd}^3 \times 1.30 = 8667 \text{ yd}^3$$

Third, determine the number of truckloads:
$$8667 \text{ yd}^3 / (12 \text{ yd}^3 / \text{truckload}) = 722 \text{ truckloads}$$

6.9 a.

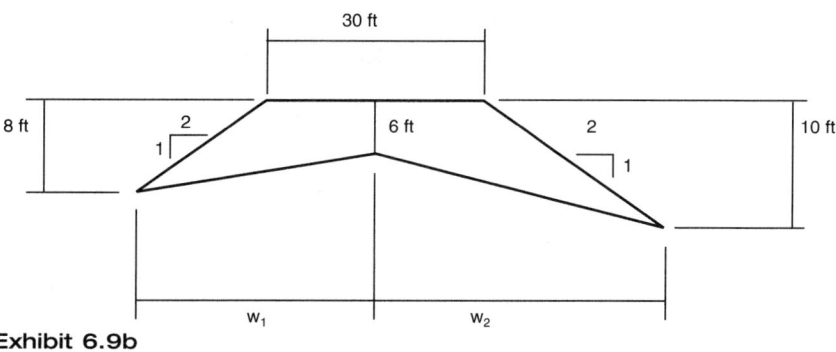

Exhibit 6.9b

$$w_1 = (30 \text{ ft})/2 + 2(8 \text{ ft}) = 31 \text{ ft}$$
$$w_2 = (30 \text{ ft})/2 + 2(10 \text{ ft}) = 35 \text{ ft}$$

Area of a triangle is one-half the product of the length of one side of the triangle and the perpendicular distance to the apex point.

$$A_A = (8 \text{ ft}) (15 \text{ ft}) / 2 = 60 \text{ ft}^2$$
$$A_B = (6 \text{ ft}) (31 \text{ ft}) / 2 = 93 \text{ ft}^2$$
$$A_C = (6 \text{ ft}) (35 \text{ ft}) / 2 = 105 \text{ ft}^2$$
$$A_D = (10 \text{ ft}) (15 \text{ ft}) / 2 = 75 \text{ ft}^2$$

$$A_{\text{total}} = A_A + A_B + A_C + A_D = 60 \text{ ft}^2 + 93 \text{ ft}^2 + 105 \text{ ft}^2 + 75 \text{ ft}^2 = 333 \text{ ft}^2$$

Alternate Solution:

$$A = \frac{h_2(w_1 + w^2)}{2} + \frac{w(h_1 + h_3)}{4}$$

$$= \frac{(6 \text{ ft})(31 \text{ ft} + 35 \text{ ft})}{2} + \frac{(30 \text{ ft})(8 \text{ ft} + 10 \text{ ft})}{4} = 333 \text{ ft}^2$$

6.10 b. First, determine the amount of soil in spoil bank:

$$V = (1/2) \, BHL \text{ (triangular spoil bank)}$$
$$V = \text{volume}$$
$$B = \text{width}$$
$$H = \text{height}$$
$$L = \text{length}$$

$$V = (1/2)(10.5 \text{ ft})(5 \text{ ft})(40 \text{ ft}) = 1050 \text{ ft}^3$$
$$V = [1050 \text{ ft}^3] \, [(1 \text{ yd}^3)/(27 \text{ ft}^3)] = 38.89 \text{ yd}^3$$

Second, determine the amount of loose soil:

$$38.89 \text{ yd}^3 \times 1.15 = 44.72 \text{ yd}^3$$

Third, determine the number of truckloads:

$$44.72 \text{ yd}^3 / (12 \text{ yd}^3 / \text{truckload}) = 3.7 \text{ truckloads}$$

Use 4 truckloads.

6.11 c. Treat the footing as a large solid block and subtract out an inner block. Do the same for the wall.

Outer dimensions of footing: $b_1 = 30 \text{ ft} + 1 \text{ ft} + 1 \text{ ft} = 32 \text{ ft}$
$h_1 = 40 \text{ ft} + 1 \text{ ft} + 1 \text{ ft} = 42 \text{ ft}$
Inside dimensions of footing: $b_2 = 30 \text{ ft} - 2 \text{ ft} - 2 \text{ ft} = 26 \text{ ft}$
$h_2 = 40 \text{ ft} - 2 \text{ ft} - 2 \text{ ft} = 36 \text{ ft}$

$$V_{\text{footing}} = [(32 \text{ ft})(42 \text{ ft}) - (26 \text{ ft})(36 \text{ ft})](1 \text{ ft}) = 408 \text{ ft}^3$$

Outer dimensions of wall: $b_3 = 30 \text{ ft}$
$h_3 = 40 \text{ ft}$

Inside dimensions of wall: $b_4 = 30 \text{ ft} - 1 \text{ ft} - 1 \text{ ft} = 28 \text{ ft}$
$h_4 = 40 \text{ ft} - 1 \text{ ft} - 1 \text{ ft} = 38 \text{ ft}$

$$V_{\text{wall}} = [(30 \text{ ft})(40 \text{ ft}) - (28 \text{ ft})(38 \text{ ft})] \, (5 \text{ ft}) = 680 \text{ ft}^3$$
$$V_{\text{total}} = V_{\text{footing}} + V_{\text{wall}} = 408 \text{ ft}^3 + 680 \text{ ft}^3 = 1088 \text{ ft}^3$$
$$V_{\text{total}} = (1088 \text{ ft}^3) \, [(1 \text{ yd}^3)/(27 \text{ ft}^3)] = 40.3 \text{ yd}^3$$
$$V_{\text{total}} = (1.05) \, (40.3 \text{ yd}^3) = 42.3$$

The answer choice that most nearly matches is **c**, 43 yd^3.

Approximate solution that is *not* exactly correct because of the corners:

$$A_{\text{cross section}} = (3' \times 1') + (1' \times 5') = 8 \text{ ft}^2$$
$$\text{Perimeter} = 40' + 30' + 40' + 30' = 140'$$
$$V_{\text{total}} = 1.05[(8 \text{ ft}^2)(140')][(1 \text{ yd}^3)/(27 \text{ ft}^3)] =$$
$$\cancel{43.6 \text{ yd}}^3 \text{ (incorrect solution)}$$

6.12 **a.** Bars running parallel to the 27' 8" dimension:
Length of rebar #1 will be 20 ft.
Length of rebar #2 = (27' 8") − (2)(2") − 20' + 2' = 9' 4".
Two 9' 4" lengths will be cut from each bar and 1' 4" will be waste.
Side dimension requiring rebar = 19' 8" − (2)(2") = 19' 4" = 19.33333 ft.
Number of spaces required = 19.33333 ft (12"/ft) / (8"/space)
 = 29 spaces.
29 + 1 = 30 rebar (Note: There is one more rebar than space).
 30 − (20 ft lengths)
 30 − (9' 4" lengths) or 15 − (20 ft lengths)
Bars running parallel to the 19' 8" dimension:
Length of rebar = (19' 8") − (2)(2") = 19' 4"
Eight inches will be cut off each bar and will be waste.
Side dimension requiring rebar = 27' 8" − (2)(2") = 27' 4" = 27.33333 ft.
Number of spaces required = 27.33333 ft (12"/ft) / (8"/space) = 41 spaces.
41 + 1 = 42 rebar (Note: There is one more rebar than space).
Total number of 20' rebars = 30 + 15 + 42 = 87 rebars.

6.13 **a.** Volume of a standard mold is 1/30 cf.
 wet density = (12.96 − 9.20) × 30 = 112.8 pcf
 moisture content = (112 g − 102 g)/102 g × 100% = 9.8%
 dry density = 112.8 pcf/1.098 = 102.7 pcf

6.14 **b.** Gallons per station = 110 pcf × (0.10 − 0.08)/100 × (32' × 100 ft × 0.5')/8.33 lb/gal = 422.6 gal/station

$$\text{Area} = (32' \times 100')/9 = 355.6 \text{ yd}^2$$
$$\text{Gal/yd}^2 = 422.6/355.6 = 1.2 \text{ gal/yd}^2$$

6.15 **c.** The soil is a Type B soil, requiring a slope of 1:1. For a depth of ten ft, this requires ten ft of extra width per side.

Total width = 4' + 10' + 10' = 24' width at top of trench excavation.

Egress ladders must also be provided every 25', and excavated material must be placed no closer than 2' from the excavation.

6.16 **c.** Bucket volume = 2.5 yd³.

Basic cycle time (from Exhibit 7.16a) = 0.35 minute
Travel time = 0.5 minute (from Exhibit 7.16b)
Total cycle time = 0.35 + 0.5 = 0.85 minute
Production = 2.5 lcy/cycle × 45 min/hr/0.85 min/cycle = 132 lcy/hr

6.17 a. Production = (5280 ft/mi × 3.5 mi/hr × 8 ft × 0.6)/(2 passes × 9 sf/sy)
= 4928 sy/hr.

On a 40 ft wide roadway, this is the equivalent of:
(4928sy/hr × 9 sf/sy)/ 40 ft width = 951 ft of road length per hr (approx. 9.5 sta.).

6.18 c. Fuel cost:
Consumption = 400 hp × 0.035 gal/hr/hp = 14 gal/hr
Cost = 14 gal/hr × $2.60/gal = $36.40/hr
Service cost = 0.33 × $36.40 = $12.00/hr

Repair cost:
Lifetime repair cost = 0.90 × ($170,000 − 10,000) = $160,000
Repair cost = 1/15 × $160,000/1000 = $10.67/hr

Tire Cost:
Cost = 1.15 × $10,000/1000 hrs = $11.50/hr

Operator's Wages:
Cost = 1.250 × $20.00 = $25.00/hr
Total operating costs = $36.40 + $12.00 + $10.67 + $11.50 + $25.00 = $95.57

6.19 c. The material Load Factor indicates their ability to take on air voids in the loose state and is obtained using the following formula:

Load Factor = pounds per cubic yards-loose/pounds per cubic yards-bank
= 2130/2960
= 0.72

Therefore,

Percent swell = [(1/0.72) − 1] × 100 = 39%
The truck complete cycle = 2.0 + 15 +1.5 + 11.5 = 30 minutes
Wheel loader complete cycle = 2.0 minutes

Since there are some periods of unproductiveness, including equipment fuelling, operator recess, and others, an efficiency adjustment must be considered.

Productivity = (45 min/hr)/(30 min/cycle) × 18 cm/cycle × 5 trucks
= 135 cm/hr

We must now determine how many loose cubic meters result from 1480 cm (compacted).

1480 cm × 1.30 = 1924

Therefore, the amount of time required to dispose of all the loose material is:

1924 cm/135 cm/hr = 14.25 hrs

6.20 d. A-B-F-G have 0 Total Float, so are on the critical path.

6.21 c. The best method to determine the total duration of a project is to draw the CPM diagram using the information provided. There are two main methods available to determine the critical path of a project: AON (Activity on Node) and AOA (Activity on Arrow). Both methods will provide the same solution; the only difference is the graphical representation of the activity. AOA could show the presence of dummy activities.

Using the AON method results in the network diagram shown in Exhibit 6.21b.

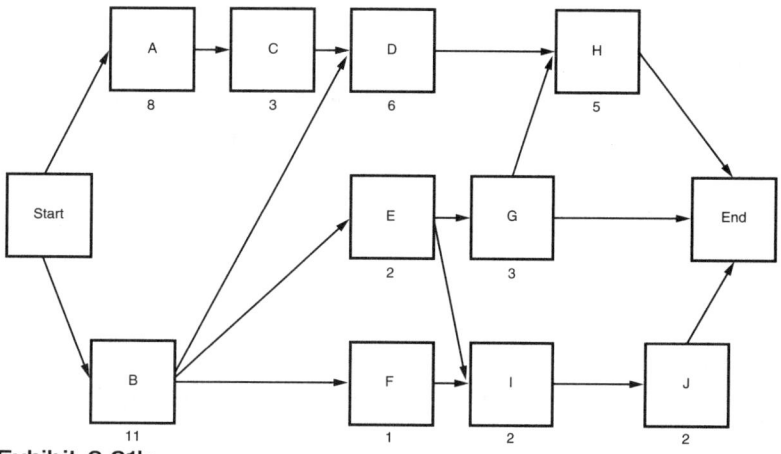

Exhibit 6.21b

Once the diagram is determined, all the sequences of activities (paths) from the beginning to the end of the project can be calculated with the correspondent duration:

ACDH – 22 Days
BDH – 22 Days
BEGH – 21 Days
BFIJ – 16 Days
BEGIJ – 20 Days

To determine the project duration, the longest path is chosen. In this example, there are two critical paths, ACDH and BDH with 22 days of duration apiece.

6.22 d. When activity duration is calculated based on a probability distribution, the Program Evaluation and Review Technique (PERT) can be used to determine the expected duration given the optimistic (a), most likely (m), and pessimistic (b) durations for each activity. The expected duration (t_e) is found as $t_e = (a + 4m + b)/6$.

Therefore, the expected duration (t_e) for each activity can be calculated and the longest path determined.

Exhibit 6.22b

Activity	Optimistic Time a (Days)	Most Likely Time m (Days)	Pessimistic Time b (Days)	Expected Time t_e (Days)
A	2	6	10	6
B	2	5	14	6
C	4	10	22	11
D	8	10	18	11
E	8	19	30	19
F	12	17	28	18
G	3	8	13	8
H	7	15	17	14
I	6	9	18	10

The following are the different paths of the project with their correspondent durations:

ACEH : 6 + 11 + 19 + 14 = 50 days
ACFI : 6 + 11 + 18 + 10 = 35 days
BDGI : 6 + 11 + 8 + 10 = 35 days

Therefore, the project expected duration (t_e) is 50 days.

6.23 b. EF time for Activity G is 23 days.

Once a list of activities with their correspondent durations and dependencies is given, a network diagram can be established. The Earliest Start (ES) of any given activity must be the same as or later than the latest of all earliest finish times of all the activities immediately preceding the given activity. Activities with no predecessor have an Early Start (ES) time of zero. The Early Finish times are found using the following formula:

EF = ES + Duration Estimate

Assuming an initial time of "zero" for the first activity, a table can be established, and the early start and finish dates are computed without reference to a diagram. This is following the rule mentioned before that the Earliest Start (ES) of any given activity must be the same as or later than the latest of all earliest finish times of all the activities immediately preceding the given activity.

Exhibit 6.23b

Activity	Duration (Days)	Inmediate Predeccesor	Early Start (ES)	Early Finish (EF)
A	5	-	0	5
B	3	-	0	3
C	6	-	0	6
D	10	A	5	15
E	12	A,B,C	6	18
F	11	C	6	17
G	5	D,E	18	23
H	7	E	18	25
I	9	F	17	26
J	10	G,H	25	35

6.24 b. The Eichleay Formula is a widely used method for calculating home office overhead damages in construction delay cases. It helps you to determine the portion of the home office overhead that should be allocated to any given project.

$$\frac{\text{Total contract price}}{\text{Total company billings}} \times \text{Total home office overhead} = \text{Allocable overhead to the delayed project}$$

$$\$7,500,000/\$25,000,000 \times \$1,500,000 = \$450,000$$

Then, a daily rate for the allocation of home office overhead must be determined:

$$\frac{\text{Allocable overhead to the delayed project}}{\text{Total duration of the project including delay days}} = \text{Overhead rate per day due to delay}$$

$$\$450,000/450 \text{ days} = \$1,000/\text{day}$$

Finally, you can calculate the home office overhead damages by multiplying the daily overhead rate due to delay by the compensable delay days:

$$\$1,000/\text{day} \times 60 \text{ days} = \$60,000$$

6.25 b. The Estimate at Completion (EAC) equals the Actual Cost of the Work Performed (ACWP) so far plus the estimated cost of the remaining work (Estimate to Complete, or ETC).

$$EAC = ACWP + ETC$$

Since the delays encountered will continue during the remaining work, the ETC should be modified by the Cost Performance Index (CPI). The CPI is found by dividing the Budgeted Cost of the Work Performed (BCWP) by the ACWP.

$$CPI = BCWP/ACWP$$

The BCWP is found by multiplying the total cost by the percentage completed.

$$BCWP = \$240{,}000 \times 60\% = \$144{,}000$$

This is the amount of money that should be invested to produce 60 percent of the work, but your actual cost is $192,000. This means that so far, you are experiencing a $48,000 cost overrun.

Calculating the CPI will tell you the adjustment that should be made to the budget of the remaining part to take into account the delay on its installation.

$$CPI = \$144{,}000 \, / \, \$192{,}000 = 0.75$$

This ratio indicates that for every $1 invested on this activity, only $0.75 of earned value is received.

The ETC, assuming that the delay will continue, is found by multiplying the original budget by the remaining percentage of the work and dividing by the CPI.

$$ETC = \$240{,}000 \times 40\% \, / \, 0.75 = \$128{,}000$$

$$EAC = \$192{,}000 + \$128{,}000 = \$320{,}000$$

Therefore, the EAC is $320,000.

6.26 c. The Schedule Variance (SV) is the difference between the budgeted cost of the work performed (BCWP) and the cost of the work to be scheduled to date (BCWS):

$$SV = BCWP - BCWS$$

The BCWP can be found by multiplying the total budget by the percentage completed:

$$BCWP = \$250{,}000 \times 65\% = \$162{,}500$$

The budgeted cost of the work that should be performed as per the schedule is the total budget times the percentage of the work that should be completed to date:

$$BCWS = \$250{,}000 \times 85\% = \$212{,}500$$

Therefore,

$$SV = \$162{,}500 - \$212{,}500 = -\$50{,}000$$

This means that given the amount of money invested, you have not accomplished as much work as you should.

6.27 a. The free float (FF), often referred as slack, of any activity is defined as the amount time the activity can be delayed without affecting the early start of its successor. Total float (TF) is the amount of time an activity can be delayed before impacting the overall project duration.

$$TF_{ij} = LF_{ij} - EF_{ij}$$

From the given diagram, the forward pass and backward pass can be calculated providing the late finish (LF) and early finish (EF) for each activity.

Therefore, the total float, or slack, for activity B is $5 - 3 = 2$. This means that activity B can be delayed a maximum of two days without becoming critical or affecting the total duration of the project.

6.28 c. Resource leveling is the method for developing a smoother distribution of resource usage. It is a trial-and-error method in which the free float, or slack, of each activity is used to maintain a uniform level of required resources. In this particular example, each task is independent. Therefore, the tasks can be arranged in any way necessary to obtain a constant usage of resources. The following activities arrangement produces a constant usage of four painters every day.

Activity	Days							
	1	2	3	4	5	6	7	8
A	■	■ 2 Painters						
B	■	■	■	■	■	■ 1 Painter		
C			■	■	■	■ 1 Painter		
D						■	■ 2 Painters	
E	■	■ 1 Painter						
F							■	■ 2 Painters
G			■	■	■	■	■	■ 2 Painters
Total Resources	4	4	4	4	4	4	4	4

Exhibit 6.28b

6.29 c. D-C-F-G was the initial critical path for the project; however, activities E and B are now behind schedule by a total of four days, and there were only two days of total float for these activities. E and B are now on the critical path, and the project is extended by two days. The new critical path is D-E-B-G.

6.30 c. Activities A and B are on schedule through day 7. Activity C has two days remaining for completion; however, there were two days of float available (days 7 and 8). One day of float (day 7) was wasted, leaving one day of float (day 8) to pick up one of the two days that Activity C is behind schedule. The other day needed for activity C will be picked up on day 9, pushing the schedule behind one day. One day will be added to the total job.

$$12 \text{ days} + 1 \text{ day} = 13 \text{ days}$$

6.31 d.

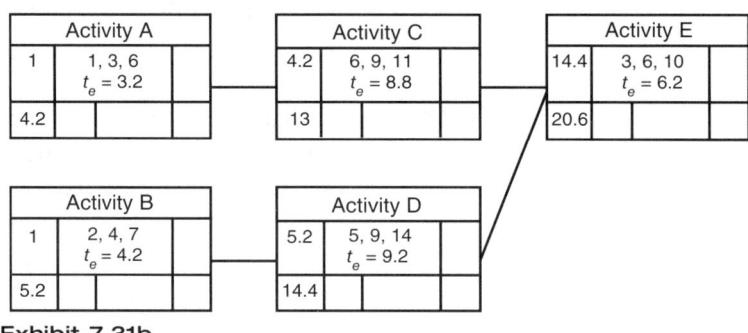

Exhibit 7.31b

$$t_e = (a + 4m + b) / 6$$
t_e = expected completion time
a = optimistic activity duration
m = most likely duration
b = pessimistic activity duration

$t_{e \text{ Activity A}} = [1 + 4(3) + 6] / 6 = 3.2$
$t_{e \text{ Activity B}} = [2 + 4(4) + 7] / 6 = 4.2$
$t_{e \text{ Activity C}} = [6 + 4(9) + 11] / 6 = 8.8$
$t_{e \text{ Activity D}} = [5 + 4(9) + 14] / 6 = 9.2$
$t_{e \text{ Activity E}} = [3 + 4(6) + 10] / 6 = 6.2$

Critical path is B-D-E. Adding the durations gives 4.2 + 9.2 + 6.2 = 19.6 days. The notation used in this question shows the start of the day on the job. The job starts on day 1. The early finish of the job is the start of day 20.6, or it is completed at the end of day 19.6; therefore the duration is 19.6 days.

6.32 a.

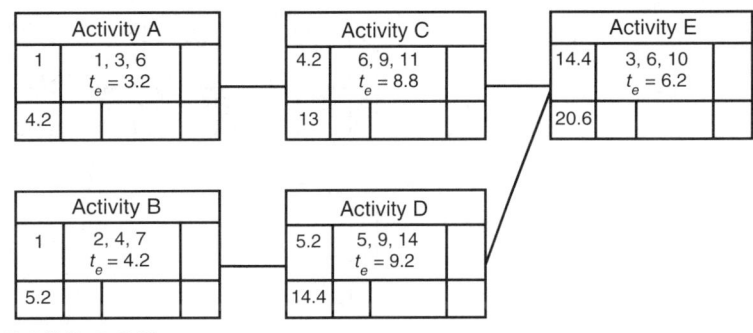

Exhibit 6.32b

$$t_e = (a + 4m + b)/6 \qquad \sigma_{te} = (b - a)/6 \qquad v = (\sigma_{te})^2$$

t_e = expected completion time
a = optimistic activity duration
m = most likely duration
b = pessimistic activity duration
σ_{te} = standard deviation
v = variance

$t_{e\ \text{Activity A}} = [1 + 4(3) + 6]/6 = 3.2 \qquad \sigma_{te} = (6 - 1)/6 = 0.833 \qquad v = (0.833)^2 = 0.694$
$t_{e\ \text{Activity B}} = [2 + 4(4) + 7]/6 = 4.2 \qquad \sigma_{te} = (7 - 2)/6 = 0.833 \qquad v = (0.833)^2 = 0.694$
$t_{e\ \text{Activity C}} = [6 + 4(9) + 11]/6 = 8.8 \qquad \sigma_{te} = (11 - 6)/6 = 0.833 \qquad v = (0.833)^2 = 0.694$
$t_{e\ \text{Activity D}} = [5 + 4(9) + 14]/6 = 9.2 \qquad \sigma_{te} = (14 - 5)/6 = 1.5 \qquad v = (1.5)^2 = 2.25$
$t_{e\ \text{Activity E}} = [3 + 4(6) + 10]/6 = 6.2 \qquad \sigma_{te} = (10 - 3)/6 = 1.167 \qquad v = (1.167)^2 = 1.361$

Critical path is B-D-E. The variance for the critical path is the sum of the variances for the activities on the critical path.

Adding the variances gives 0.694 + 2.25 + 1.361 = 4.305 days → 4.3 days

6.33 b.

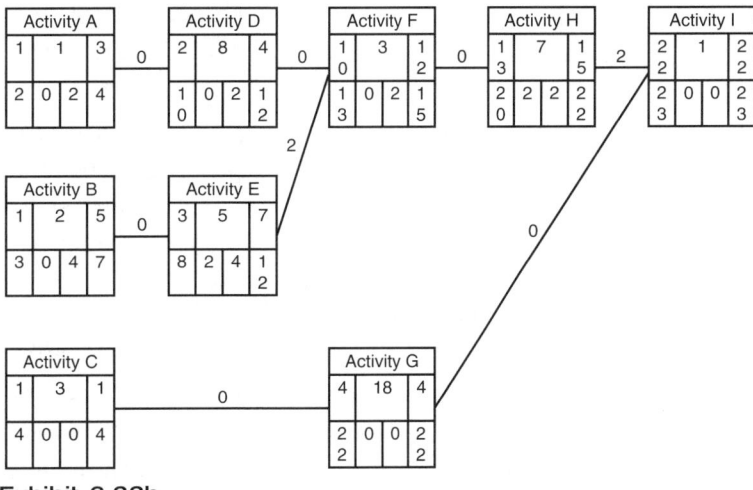

Exhibit 6.33b

6.34 a.

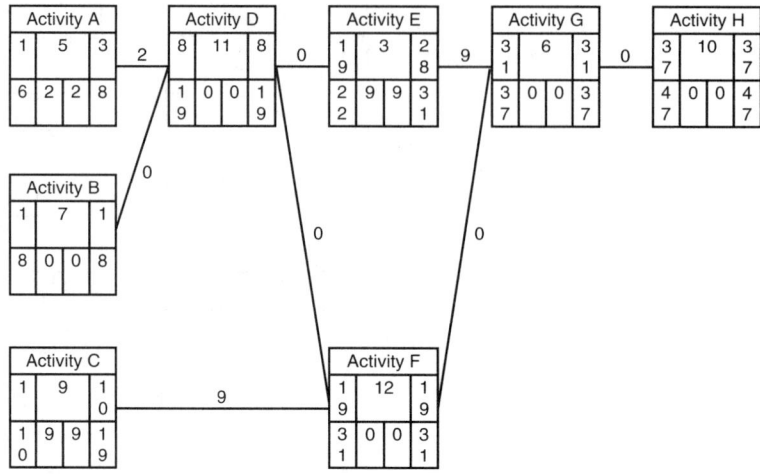

Exhibit 6.34b

The critical path consists of those activities along the line from beginning to end of project that has zero floats.

6.35 b. Critical path is B-D-F-G-H. Adding the durations gives 7 + 11 + 12 + 6 + 10 = 46 days. The notation that was used in question 7.34 shows the start of the day on the job. The job starts on day 1. The early finish of the job is the start of day 47, or it is completed at the end of day 46; therefore, the duration is 46 days.

6.36 b.

$$F = P(1 + i)^n$$
F = Future value
P = Present value
i = Interest rate
n = Number of interest periods

$P_{12/2003} = \$700{,}000$

$F_{12/2004} = \$700{,}000 \, (1 + 0.055)^1 = \$738{,}500$

$F_{12/2005} = \$738{,}500 \, (1 + 0.04)^1 = \$768{,}040$

$F_{12/2006} = \$768{,}040 \, (1 + 0.05)^1 = \$806{,}442$

$F_{12/2007} = \$806{,}442 \, (1 + 0.05)^1 = \$846{,}764.10$

$F_{5/2008} = \$846{,}764.10 + (5/12)[\$846{,}764.10 \, (0.06)] =$ $\$867{,}933.20$, which is nearest to answer choice **b**, $\$868{,}000$

6.37 a.

$$D_j = (2C/m)[1 - (2/n)]^{j-1}$$
$$BV_j = C \, [1 - (2/n)]^j$$
D_j = Depreciation for year j
n = Number of years of useful life

C = Initial cost
S_n = Salvage value
BV = Book value
j = Year that is evaluated
$BV_3 = \$150,000\,[1-(2/5)]^3 = \$32,400$ (Note: If this value is below the salvage value, then use salvage value.)

Alternate solution #1:

$D_1 = [(2 \times \$150,000)/5][1 - (2/5)]^{1-1} = \$60,000$
$D_2 = [(2 \times \$150,000)/5]\,[1 - (2/5)]^{2-1} = \$36,000$
$D_3 = [(2 \times \$150,000)/5]\,[1 - (2/5)]^{3-1} = \$21,600$
$BV_3 = \$150,000 - (\$60,000 + \$36,000 + \$21,600) = \$32,400$ (Note: If this value is below the salvage value, then use salvage value.)

Alternate solution #2:

Double declining balance method is a constant percentage method.
Yearly percentage = 2(100%)/5 = 40%
$D_1 = 40\%\,(\$150,000) = \$60,000$
$BV_1 = \$150,000 - \$60,000 = \$90,000$
$D_2 = 40\%\,(\$90,000) = \$36,000$
$BV_2 = \$90,000 - \$36,000 = \$54,000$
$D_3 = 40\%\,(\$54,000) = \$21,600$
$BV_3 = \$54,000 - \$21,600 = \$32,400$ (Note: if this value is below the salvage value, then use salvage value.)

6.38 d.

$$D_j = \frac{n-j+1}{(n/2)(n+1)}(C-S_n)$$

$$BV = C - \Sigma(D_j)$$

D_j = Depreciation for year j
j = Year that is evaluated
n = Number of years of useful life
C = Initial cost
S_n = Salvage value
BV = Book value

$D_1 = (4 - 1 + 1)/[(4/2)(4 + 1)](\$38,000 - \$7,000) = \$12,400$
$D_2 = (4 - 2 + 1)/[(4/2)(4 + 1)](\$38,000 - \$7,000) = \$9,300$
$D_3 = (4 - 3 + 1)/[(4/2)(4 + 1)](\$38,000 - \$7,000) = \$6,200$
$BV_3 = \$38,000 - (\$12,400 + \$9,300 + \$6,200) = \$10,100$

Alternate solution:
Sum = 1 + 2 + 3 + 4 = 10
D_1 = (4/10) ($38,000 − $7,000) = $12,400
D_2 = (3/10) ($38,000 − $7,000) = $9,300
D_3 = (2/10) ($38,000 − $7,000) = $6,200
D_4 = (1/10) ($38,000 − $7,000) = $3,100
BV_3 = $38,000 − ($12,400 + $9,300 + $6,200) = $10,100

Note: $D_1 + D_2 + D_3 + D_4$ = $12,400 + $9,300 + $6,200 + $3,100 = $31,000, which should be same as ($38,000 − $7,000) = $31,000.

6.39 d.

$$\text{Payment} = \text{Amount borrowed} \times 4 \times \frac{1-(1+i)^{-n}}{i}$$

i = interest rate for the payment period
n = number of payments

i = 0.07/12 = 0.00583333333
n = 5 × 12 = 60

$$\text{Payment} = (\$80{,}000 - \$5000) \times 4 \times \frac{1-(1+0.00583333333)^{-60}}{0.00583333333}$$

Payment = $1485.09

6.40 c.

$$\text{Final amount} = \text{monthly payment} \times \frac{(1+i)^n - 1}{i}$$

i = interest rate per deposit period = (yearly interest rate)/12
n = total number of deposits

i = 0.03/12 = 0.0025
n = (15 years) (12 months/year) = 180 months

$$\text{Final amount} = \$500 \times \frac{(1+0.0025)^{180} - 1}{0.0025} = \$113{,}486.34 \approx \$113{,}500$$

SOLUTION SUMMARIES

Solution Summary for Chapter 1 Breadth Exam

Problem	Solution
1.1	c.
1.2	b.
1.3	c.
1.4	a.
1.5	c.
1.6	c.
1.7	d.
1.8	a.
1.9	b.
1.10	d.
1.11	c.
1.12	c.
1.13	d.
1.14	c.
1.15	b.
1.16	d.
1.17	b.
1.18	c.
1.19	c.
1.20	b.
1.21	d.
1.22	b.
1.23	c.
1.24	c.
1.25	c.
1.26	b.
1.27	b.
1.28	b.
1.29	b.
1.30	d.
1.31	c.
1.32	b.
1.33	c.
1.34	c.
1.35	b.
1.36	d.
1.37	d.
1.38	c.
1.39	b.
1.40	b.

Solution Summary for Chapter 2 Environmental Engineering and Water Resources Exam

Problem	Solution
2.1	a.
2.2	c.
2.3	c.
2.4	a.
2.5	d.
2.6	b.
2.7	a.
2.8	c.
2.9	b.
2.10	d.
2.11	b.
2.12	b.
2.13	c.
2.14	c.
2.15	b.
2.16	d.
2.17	a.
2.18	a.
2.19	b.
2.20	c.
2.21	d.
2.22	a.
2.23	b.
2.24	b.
2.25	d.
2.26	b.
2.27	c.
2.28	b.
2.29	c.
2.30	b.
2.31	a.
2.32	d.
2.33	b.
2.34	b.
2.35	b.
2.36	d.
2.37	d.
2.38	c.
2.39	a.
2.40	a.

Solution Summary for Chapter 3 Geotechnical Engineering Exam

Problem	Solution
3.1	b.
3.2	b.
3.3	c.
3.4	b.
3.5	c.
3.6	c.
3.7	c.
3.8	b.
3.9	c.
3.10	c.
3.11	c.
3.12	c.
3.13	c.
3.14	b.
3.15	a.
3.16	b.
3.17	c.
3.18	b.
3.19	a.
3.20	c.
3.21	b.
3.22	a.
3.23	c.
3.24	c.
3.25	c.
3.26	b.
3.27	b.
3.28	c.
3.29	a.
3.30	b.
3.31	c.
3.32	b.
3.33	b.
3.34	d.
3.35	c.
3.36	d.
3.37	a.
3.38	b.
3.39	d.
3.40	d.

Solution Summary for Chapter 4 Structural Engineering Exam

Problem	Solution
4.1	c.
4.2	b.
4.3	d.
4.4	a.
4.5	c.
4.6	a.
4.7	c.
4.8	c.
4.9	a.
4.10	c.
4.11	d.
4.12	c.
4.13	b.
4.14	b.
4.15	b.
4.16	b.
4.17	c.
4.18	c.
4.19	a.
4.20	c.
4.21	c.
4.22	d.
4.23	a.
4.24	c.
4.25	c.
4.26	d.
4.27	c.
4.28	c.
4.29	b.
4.30	d.
4.31	b.
4.32	c.
4.33	d.
4.34	d.
4.35	d.
4.36	b.
4.37	d.
4.38	a.
4.39	c.
4.40	c.

Solution Summary for Chapter 5 Transportation Engineering Exam

Problem	Solution
5.1	d.
5.2	d.
5.3	c.
5.4	d.
5.5	d.
5.6	b.
5.7	d.
5.8	a.
5.9	c.
5.10	d.
5.11	d.
5.12	b.
5.13	c.
5.14	c.
5.15	b.
5.16	a.
5.17	c.
5.18	b.
5.19	d.
5.20	b.
5.21	c.
5.22	b.
5.23	c.
5.24	b.
5.25	a.
5.26	c.
5.27	c.
5.28	a.
5.29	c.
5.30	c.
5.31	d.
5.32	a.
5.33	c.
5.34	d.
5.35	a.
5.36	c.
5.37	a.
5.38	c.
5.39	c.
5.40	d.

Solution Summary for Chapter 6 Construction Engineering Exam

Problem	Solution
6.1	c.
6.2	b.
6.3	c.
6.4	b.
6.5	c.
6.6	a.
6.7	a.
6.8	d.
6.9	a.
6.10	b.
6.11	c.
6.12	a.
6.13	a.
6.14	b.
6.15	c.
6.16	c.
6.17	a.
6.18	c.
6.19	c.
6.20	d.
6.21	c.
6.22	d.
6.23	b.
6.24	b.
6.25	b.
6.26	c.
6.27	a.
6.28	c.
6.29	c.
6.30	c.
6.31	d.
6.32	a.
6.33	b.
6.34	a.
6.35	b.
6.36	b.
6.37	a.
6.38	d.
6.39	d.
6.40	c.